Lecture Notes in Computer Science

Commenced Publication in 1973
Founding and Former Series Editors:
Gerhard Goos, Juris Hartmanis, and Jan van Leeuwen

Rudolf Eigenmann
Bronis R. de Supinski (Eds.)

OpenMP in a New Era of Parallelism

4th International Workshop, IWOMP 2008
West Lafayette, IN, USA, May 12-14, 2008
Proceedings

 Springer

Volume Editors

Rudolf Eigenmann
Purdue University
School of Electrical and Computer Engineering
Electrical Engineering Building
465 Northwestern Ave.
West Lafayette, IN 47907-2035, USA
E-mail: eigenman@purdue.edu

Bronis R. de Supinski
Computation Directorate
Lawrence Livermore National Laboratory
Livermore, CA 94551, USA
E-mail: bronis@llnl.gov

Library of Congress Control Number: 2008925752

CR Subject Classification (1998): D.1.3, D.1, D.2, F.2, G.1-4, J.2, I.6

LNCS Sublibrary: SL 1 – Theoretical Computer Science and General Issues

ISSN   0302-9743
ISBN-10   3-540-79560-X Springer Berlin Heidelberg New York
ISBN-13   978-3-540-79560-5 Springer Berlin Heidelberg New York

Springer is a part of Springer Science+Business Media

springer.com

© Springer-Verlag Berlin Heidelberg 2008

Typesetting: Camera-ready by author, data conversion by Scientific Publishing Services, Chennai, India
Printed on acid-free paper   SPIN: 12264346   06/3180   5 4 3 2 1 0

Preface

OpenMP is a widely accepted, standard application programming interface (API) for high-level shared-memory parallel programming in Fortran, C, and C++. Since its introduction in 1997, OpenMP has gained support from most high-performance compiler and hardware vendors. Under the direction of the OpenMP Architecture Review Board (ARB), the OpenMP specification has evolved, including the recent release of Specification 3.0. Active research in OpenMP compilers, runtime systems, tools, and environments drives its evolution, including new features such as tasking.

The community of OpenMP researchers and developers in academia and industry is united under cOMPunity (www.compunity.org). This orgainaization has held workshops on OpenMP around the world since 1999: the European Workshop on OpenMP (EWOMP), the North American Workshop on OpenMP Applications and Tools (WOMPAT), and the Asian Workshop on OpenMP Experiences and Implementation (WOMPEI) attracted annual audiences from academia and industry. The International Workshop on OpenMP (IWOMP) consolidated these three workshop series into a single annual international event that rotates across the previous workshop sites. The first IWOMP meeting was held in 2005, in Eugene, Oregon, USA. IWOMP 2006 took place in Reims, France, and IWOMP 2007 in Beijing, China. Each workshop drew over 60 participants from research and industry throughout the world. IWOMP 2008 continued the series with technical papers, panels, tutorials, and OpenMP status reports. The first IWOMP workshop was organized under the auspices of cOMPunity. Since that workshop, the IWOMP Steering Committee has organized these events and guided development of the series. The first three IWOMP meetings were successful in every regard, due largely to the generous support received from numerous sponsors, demonstrating the importance of OpenMP as a practical programming paradigm.

The cOMPunity website (www.compunity.org) provides access to the talks given at the meetings and to photos of the activities. The IWOMP website (www.iwomp.org) provides information on the latest event. This book contains the proceedings of IWOMP 2008. The workshop program included 16 technical papers, a keynote talk on "Programming with Transactions" by Kunle Olukotun, Stanford University, and an invited talk entitled "Taking OpenMP Beyond HPC" by Tim Mattson, Intel Corp. The workshop program also featured a panel entitled "Is OpenMP Irrelevant Next to MPI and Pthreads?"—of course, the title created lively discussions, providing evidence to the contrary! The workshop concluded with a status and update report by OpenMP's driving force—its ARB—and several OpenMP vendors.

Rudi Eigenmann
Bronis R. de Supinski

Organization

Committee

General Chair	Rudi Eigenmann (Purdue University, USA)
Publicity Chair	Matthijs van Waveren (Fujitsu, France)
Publications Chair	Bronis R. de Supinski (LLNL, USA)
Publications Chair	Eduard Ayguadé (Barcelona Supercomputing Centre, Spain)
Sponsors Contact Chair	Barbara Chapman (University of Houston, USA)
Tutorials Chair	Ruud van der Pas (Sun Microsystems, Netherlands)
Website Chair	Chirag Dave (Purdue University, USA)
Local Arrangements Chair	Melanie A. Lindsay (Purdue University, USA)
Program Committee Chair	Rudi Eigenmann (Purdue University, USA)
Program Committee	Dieter an Mey (RWTH Aachen University, Germany)
	Eduard Ayguadé (Barcelona Supercomputing Center, Spain)
	Barbara Chapman (University of Houston, USA)
	Bronis R. de Supinski (LLNL, USA)
	Guang R. Gao (University of Delaware, USA)
	Rick Kufrin (NCSA/University of Illinois, USA)
	Federico Massaioli (CASPUR, Italy)
	Lawrence Meadows (Intel, USA)
	Matthias S. Müller (ZIH, TU Dresden, Germany)
	Mitsuhisa Sato (University of Tsukuba, Japan)
	Ruud van der Pas (Sun Microsystems, Netherlands)

IWOMP Steering Committee

Chair	Bronis R. de Supinski (LLNL, USA)
Committee Members	Dieter an Mey (RWTH Aachen University, Germany)

Committee Members

Dieter an Mey (RWTH Aachen University,
  Germany)
Eduard Ayguadé (Barcelona Supercomputing
  Center, Spain)
Mark Bull (EPCC, UK)
Barbara Chapman (CEO of cOMPunity,
  University of Houston, USA)
Rudi Eigenmann (Purdue University, USA)
Guang R. Gao (University of Delaware, USA)
Ricky Kendall (Oak Ridge National
  Laboratory, USA)
Michaël Krajecki (University of Reims, France)
Rick Kufrin (NCSA/University of Illinois,
  USA)
Federico Massaioli (CASPUR, Italy)
Lawrence Meadows (Intel, USA)
Matthias S. Müller (ZIH, TU Dresden,
  Germany)
Arnaud Renard (University of Reims, France)
Mitsuhisa Sato (University of Tsukuba, Japan)
Sanjiv Shah (Intel, USA)
Ruud van der Pas (Sun Microsystems,
  Netherlands)
Matthijs van Waveren (Fujitsu, France)
Michael Wong (IBM, Canada)
Weimin Zheng (Tsinghua University, China)

Table of Contents

OpenMP Tasking Models and Extensions

Applications, Scheduling, Tools

A Microbenchmark Study of OpenMP Overheads under Nested Parallelism

Vassilios V. Dimakopoulos, Panagiotis E. Hadjidoukas,
and Giorgos Ch. Philos

Department of Computer Science, University of Ioannina
Ioannina, Greece, GR-45110
{dimako,phadjido,gfilos}@cs.uoi.gr

Abstract. In this work we present a microbenchmark methodology for assessing the overheads associated with nested parallelism in OpenMP. Our techniques are based on extensions to the well known EPCC microbenchmark suite that allow measuring the overheads of OpenMP constructs when they are effected in inner levels of parallelism. The methodology is simple but powerful enough and has enabled us to gain interesting insight into problems related to implementing and supporting nested parallelism. We measure and compare a number of commercial and freeware compilation systems. Our general conclusion is that while nested parallelism is fortunately supported by many current implementations, the performance of this support is rather problematic. There seem to exist issues which have not yet been addressed effectively, as most OpenMP systems do not exhibit a graceful reaction when made to execute inner levels of concurrency.

1 Introduction

OpenMP [1] has become a standard paradigm for shared memory programming, as it offers the advantage of simple and incremental parallel program development, in a high abstraction level. Nested parallelism has been a major feature of OpenMP since its very beginning. As a programming style, it provides an elegant solution for a wide class of parallel applications, with the potential to achieve substantial processor utilization, in situations where outer-loop parallelism simply can not. Despite its significance, nested parallelism support was slow to find its way into OpenMP implementations, commercial and research ones alike. Even nowadays, the level of support is varying greatly among compilers and runtime systems.

For applications that have enough (and balanced) outer-loop parallelism, a small number of coarse threads is usually enough to produce satisfactory speedups. In many other cases though, including situations with multiple nested loops, or recursive and irregular parallel applications, threads should be able to create new teams of threads because only a large number of threads has the potential to achieve good utilization of the computational resources.

R. Eigenmann and B.R. de Supinski (Eds.): IWOMP 2008, LNCS 5004, pp. 1–12, 2008.
© Springer-Verlag Berlin Heidelberg 2008

Although many contemporary OpenMP compilation systems provide some kind of support for nested parallelism, there has been no evaluation of the overheads incurred by such a support. The well known EPCC microbenchmark suite [2,3] is a valuable tool with the ability to reveal various synchronization and scheduling overheads, but only for single-level parallelism.

In this work, we present a set of benchmarks that are based on extensions to the EPCC microbenchmarks and allow us to measure the overheads of OpenMP systems when nested parallelism is in effect. To the best of our knowledge this is the first study of its kind as all others have been based on application speedups [4,5,6,7] which give overall performance indications but do not reveal potential construct-specific problems.

The paper is organized as follows. In Section 2 we give an overview of the OpenMP specification and the current status of various implementations with respect to nested parallelism. In Section 3 we present the microbenchmarks in detail. Section 4 reports on the performance of several OpenMP compilation systems when used to execute our benchmarks. The section also includes a discussion of our findings. Finally, Section 5 concludes this work.

2   Nested Parallelism in OpenMP

The OpenMP specification leaves support for nested parallelism as optional, allowing an implementation to serialize the nested parallel region, i.e. execute it by only 1 thread. In implementations that support nested parallelism, the user can choose to enable or disable it either during program startup through the `OMP_NESTED` environmental variable or dynamically at runtime through an `omp_set_nested()` call. The number of threads that will comprise a team can be controlled by the `omp_set_num_threads()` call. Because this is allowed to appear only in sequential regions of the code, there is no way to specify a different number of threads for inner levels through this call; to overcome this, the current version of OpenMP (2.5) provides the `num_threads(n)` clause. Such a clause can appear in a (nested) `parallel` directive and request that this particular region be executed by exactly `n` threads.

However, the actual number of threads dispatched in a (nested) `parallel` region depends also on other things. OpenMP provides a mechanism for the *dynamic adjustment* of the number of threads which, if activated, allows the implementation to spawn fewer threads than what is specified by the user. In addition to dynamic adjustment, factors that may affect the actual number of threads include the nesting level of the region, the support/activation of nested parallelism and the peculiarities of the implementation. For example, some systems maintain a fixed pool of threads, usually equal in size to the number of available processors. Nested parallelism is supported as long as free threads exist in the pool, otherwise it is dynamically disabled. As a result, a nested `parallel` region may be executed by a varying number of threads, depending on the current state of the pool.

In general, it is a recognized fact that the current version of OpenMP has a number of shortcomings when it comes to nested parallelism [7], and there exist issues which need clarification. Some of them are settled in the upcoming version of the API (3.0), which will also offers a richer functional API for the application programmer.

According to the OpenMP specification, an implementation which serializes nested `parallel` regions, even if nested parallelism is enabled by the user, is considered *compliant*. An implementation can claim *support* of nested parallelism if nested `parallel` regions may be executed by more than 1 thread. Because of the difficulty in handling efficiently a possibly large number of threads, many implementations provide support for nested parallelism but with certain limitations. For example, there exist systems that support a fixed number of nesting levels; some others allow an unlimited number of nesting levels but have a fixed number of simultaneously active threads.

Regarding proprietary compilers, not all of them support nested parallelism and some support it only in part. Among the ones that provide unlimited support in their recent releases are the Fujitsu PRIMEPOWER compilers, the HP compilers for the HP-UX 11i operating system, the Intel compilers [8] and the Sun Studio compilers [9]. Full support for nested parallelism is also provided in the latest version of the well-known open-source GNU Compiler Collection, GCC 4.2, through the libGOMP [10] runtime library.

Research/experimental OpenMP compilers and runtime systems that support nested parallelism include MaGOMP, a port of libGOMP on top of the Marcel threading library [11], the Omni compiler [12,6] and OMPi [13,14].

3   The Microbenchmark Methodology

The EPCC microbenchmark suite [2,3] is the most commonly used tool for measuring runtime overheads of individual OpenMP constructs. However, it is only applicable to single-level parallelism. This section describes the extensions we have introduced to this microbenchmark suite for the evaluation of OpenMP runtime support under nested parallelism.

The technique used to measure the overhead of OpenMP directives, is to compare the time taken for a section of code executed sequentially with the time taken for the same code executed in parallel, enclosed in a given directive. Let T_p be the execution time of a program on p processors and T_1 be the execution time of its sequential version. The overhead of the parallel execution is defined as the total time spent collectively by the p processors over and above T_1, the time required to do the "real" work, i.e. $T_{ovh} = pT_p - T_1$. The per-processor overhead is then $T_o = T_p - T_1/p$. The EPCC microbenchmarks [2] measure T_o for the case of single-level parallelism using the method described below.

A reference time, T_r, is first fixed, which represents the time needed for a call to a particular function named `delay()`. To avoid measuring times that are smaller than the clock resolution, T_r is actually calculated by calling the `delay()` function sufficiently many times:

```
for (j = 0; j < innerreps; j++)
   delay(delaylength);
```

and dividing the total time by `innerreps`.

Then, the same function call is surrounded by the OpenMP construct under measurement, which in turn is enclosed within a `parallel` directive. For example, the `testfor()` routine that measures the `for` directive overheads, actually measures the portion shown in Fig. 1 and then divides it by `innerreps`, obtaining T_p. Notice, that because the measurement includes the time taken by the `parallel` directive, `innerreps` is large enough so that the overhead of the enclosing `parallel` directive can be ignored. The overhead is derived as $T_p - T_r$, since the total work done needs actually pT_r sequential time. Of course, to obtain statistically meaningful results, each overhead measurement is repeated several times and the mean and standard deviation are computed over all measurements. This way, the microbenchmark suite neither requires exclusive access to a given machine nor is seriously affected by background processes in the system.

```
testfor() {
   ...
   <start measurement>
     #pragma omp parallel private(j)
     {
       for (j = 0; j < innerreps; j++)
         #pragma omp for
           for (i = 0; i < p; i++)
             delay(delaylength);
     }
   <stop measurement>
   ...
}
```

Fig. 1. Portion of the `testfor()` microbenchmark routine

3.1   Extensions for Nested Parallelism

To study how efficiently OpenMP implementations support nested parallelism, we have extended both the synchronization and the scheduling microbenchmarks of the EPCC suite. According to our approach, the core benchmark routine for a given construct (e.g. the `testfor()` discussed above) is represented by a *task*. Each task has a unique identifier and utilizes its own memory space for storing its table of runtime measurements. We create a team of threads, where each member of the team executes its own task. When all tasks finish, we measure their total execution time and compute the global mean of all measured runtime overheads. Our approach is outlined in Fig. 2. The team of threads that execute the tasks expresses the outer level of parallelism, while each benchmark routine (task) contains the inner level of parallelism.

```
      void nested_benchmark(char *name, func_t originalfunc) {
      int     task_id;
      double t0, t1;

1     t0 = getclock();
2     #ifdef NESTED_PARALLELISM
3     #pragma omp parallel for schedule(static,1)
4     #endif
5     for (task_id = 0; task_id < p; task_id++) {
6        (*originalfunc)(task_id);
7     }
8     t1 = getclock();

      <compute global statistics>
      <print construct name, elapsed time (t1-t0), statistics>
      }


      main() {
        <compute reference time>
        omp_set_num_threads(omp_get_num_procs());
        omp_set_dynamic(0);
        nested_benchmark("PARALLEL", testpr);
        nested_benchmark("FOR",      testfor);
        ...
      }
```

Fig. 2. Extended microbenchmarks for nested parallelism overhead measurements

In Fig. 2, if the outer loop (lines 5–7) is not parallelized, the tasks are executed in sequential order. This is equivalent to the original version of the microbench-marks, having each core benchmark repeated more than once. On the other hand, if nested parallelism is enabled, the loop is parallelized (lines 2–4) and the tasks are executed in parallel. The number of simultaneously active tasks is bound by the number of OpenMP threads that constitute the team of the first level of parallelism. To ensure that each member of the team executes exactly one task, a static schedule with chunksize of 1 was chosen at line 3. In addition, to guarantee that the OpenMP runtime library does not assign fewer threads to inner levels than in the outer one, dynamic adjustment of threads is *disabled* through a call to omp_set_dynamic(0).

By measuring the aggregated execution time of the tasks, we use the mi-crobenchmark as an individual application. This time does not only include the parallel portion of the tasks, i.e. the time the tasks spend on measuring the runtime overhead, but also their sequential portion. This means that even if the mean overhead increases when tasks are executed in parallel, as expected due to the higher number of running threads, the overall execution time may decrease.

In OpenMP implementations that provide full nested parallelism support, inner levels spawn more threads than the number of physical processors, which are mostly kernel-level threads. Thus, measurements exhibit higher variations than in the case of single-level parallelism. In addition, due to the presence of more than one team parents, the overhead of the parallel directive increases in most implementations, possibly causing overestimation of other measured overheads (see Fig. 1). To resolve these issues, we increase the number of internal repetitions (`innerreps`) for each microbenchmark, so as to be able to reach the same confidence levels (95%). A final subtle point is that when the machine is oversubscribed, each processor will be timeshared among multiple threads. This leads to an overestimation of the overheads because the microbenchmarks account for the sequential work (T_r) multiple times. We overcame this by decreasing `delaylength` so that T_r becomes negligible with respect to the measured overhead.

4   Results

All our measurements were taken on a Compaq Proliant ML570 server with 4 Intel Xeon III single-core CPUs running Debian Linux (2.6.6). Although this is a relatively small SMP machine, size is not an issue here. Our purpose is to create a significant number of threads; as long as a lot more threads than the available processors are active, the desired effect is achieved. We provide performance results for two free commercial and three freeware OpenMP C compilers that support nested parallelism. The commercial compilers are the Intel C++ 10.0 compiler (ICC) and Sun Studio 12 (SUNCC) for Linux. The freeware ones are GCC 4.2.0, Omni 1.6 and OMPi 0.9.0. As Omni and OMPi are source-to-source compilers, we have used GCC as the native back-end compiler for both of them. In addition, because OMPi is available with a multitude of threading libraries, we have used two different configurations for it, namely OMPi+POSIX and OMPi+PSTHREADS. The first one uses the default runtime library, based on POSIX threads which, although optimized for single-level parallelism, provides basic support for nested parallelism. The second one uses a high-performance runtime library based on POSIX threads and portable user-level threads [14].

Most implementations start by creating an initial pool of threads, usually equal in size to the number of available processors, which is 4 in our case. Because the number of threads in the second level is implementation dependent, in all our experiments we have explicitly set it to 4 through an `omp_set_num_threads(4)` call and we have disabled the dynamic adjustment of the number of threads. I.e., when executing the second level of parallelism, there are in total $4 \times 4 = 16$ active threads. However, some implementations cannot handle this situation. In particular, the Omni compiler and OMPi+POSIX cannot create more threads on the fly, even if needed; they support nested parallelism as long as the initial pool has idle threads, otherwise nested parallel regions get serialized. To overcome this problem, for those two implementations we set the `OMP_NUM_THREADS` environmental variable equal to 16 before executing the benchmarks, so that the

initial pool is forced to have 16 threads; the `omp_set_num_threads`(4) call then
limits the outer level to exactly 4 threads, while all 16 threads are utilized in the
inner level. We have, however, been careful not to give those two implementations the advantage of zero thread creation overhead (since with the above trick
the 16 threads are pre-created), by including a dummy nested parallel region
at the top of the code. This way, all implementations get a chance to create 16
threads before the actual measurements commence.

A selection of the obtained results is given in Figs 2–4, for the synchronization
and scheduling microbenchmarks. Fig. 3 includes the overheads of all six systems
for the `parallel`, `for`, `single` and `critical` constructs. Each plot includes the
single-level overheads of each system for reference.

As the number of active threads increases when nested parallelism is enabled, the overheads are expected to increase accordingly. We observe, however,
that the `parallel` construct does not scale well for the Intel, GCC and Omni
compilers, although ICC remains quite fast. For all three of them, the runtime
overhead is more than an order of magnitude higher in the case of nested parallelism. For ICC this could be attributed, in part, to the fact that threads join a
unique central pool before getting grouped to teams [8]. On the other hand, both
OMPi+PSTHREADS and SUNCC clearly scale better and their overheads increase
linearly, with SUNCC, however, exhibiting higher overheads than OMPi for both
single level and nested parallelism.

Similar behavior is seen for the `for` and `single` constructs, except that GCC
shows significant but not excessive increase. The Sun compiler seems to handle loop scheduling quite well showing a decrease in the actual overheads. This,
combined with the decrease in the `single` overheads, reveals efficient team management since both constructs incur mostly inter-team contention. Especially
in the `single` construct, OMPi+PSTHREADS shows the advantage of user-level
threading: inner levels are executed by user-level threads, which mostly live in
the processor where the parent thread is, eliminating most inter-team contention
and the associated overheads. In contrast, the (unnamed) `critical` construct
incurs global contention since all threads from all teams must compete for a
single lock protecting the critical code section. Overheads are increased significantly in all systems, suggesting that *unnamed* `critical` constructs should be
avoided when nested parallelism is required.

Fig. 4 includes results from the scheduling microbenchmarks. For presentation
clarity, we avoided reporting curves for a wide range of chunksizes; instead, we
include only results for static, dynamic and guided schedules with a chunksize of
1, which represent the worst cases, with the highest possible scheduling overhead.
Scheduling overheads increase, as expected, for the static and guided schedules
in the case of nested parallelism. However, the overheads of the dynamic scheduling policy seem to increase at a slower rate and in some cases (SUNCC, GCC and
OMPi+PSTHREADS) actually decrease, which seems rather surprising. This can
be explained by the fact that for this particular scheduling strategy and with this
particular chunk size, the overheads are dominated by the excessive contention
among the participating threads. With locality-biased team management, which

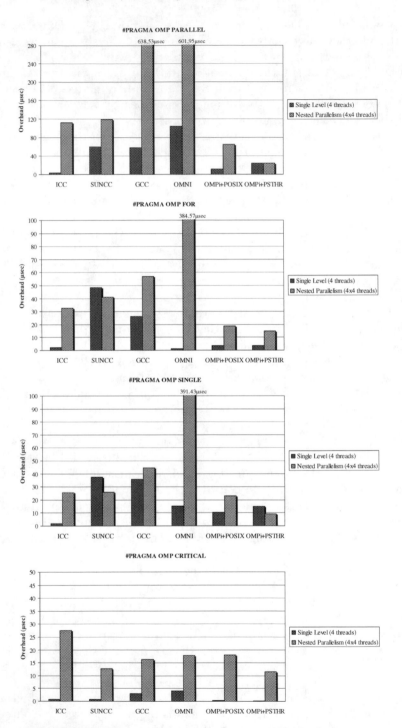

Fig. 3. Overheads for `parallel`, `for`, `single` and `critical`

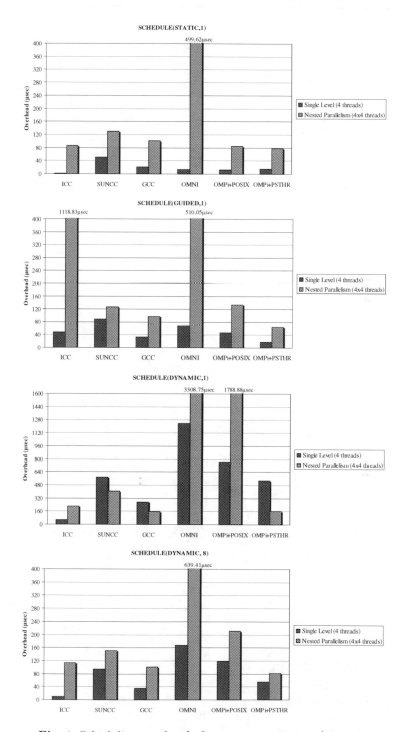

Fig. 4. Scheduling overheads for static, guided and dynamic

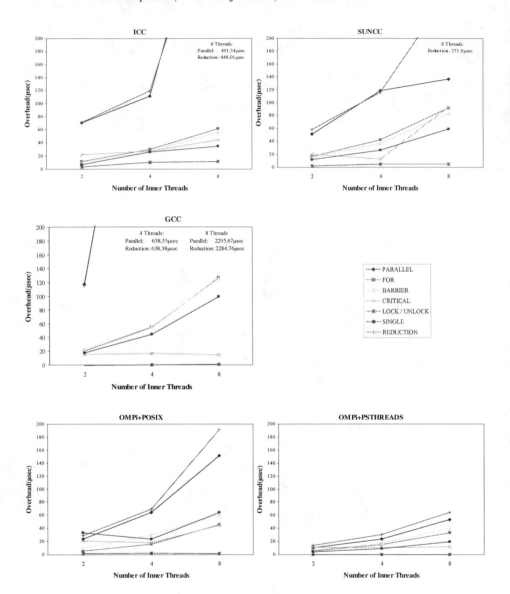

Fig. 5. Overheads per compiler, for increasing team sizes at the second level of parallelism

groups all team threads onto the same CPU, and efficient locking mechanisms, which avoid busy waiting, the contention has the potential to drop sharply, yielding lower overheads than in the single-level case. This appears to be the case for the Sun Studio and GCC compilers. OMPi with user-level threading achieves the same goal because it is able to assign each independent loop to a team of non-preemptive user-level OpenMP threads that mainly run on the same processor. However, as the chunksize increases, assigned jobs become coarser and any gains due to contention avoidance vanish. This is confirmed in the last plot of Fig. 4; with a chunksize of 8 all implementations show increased overheads with respect to the single-level case.

In Fig. 5 we present the results of our next experimentation: we delved into discovering how the behavior of our subjects changes for different populations of threads. We fixed the number of first-level threads to 4 but changed the second-level teams to consist of 2, 4 and 8 threads, yielding in total 8, 16 and 32 threads on the 4 processors. Because this was only possible using the `num_threads()` clause (an OpenMP V.2.0 addition), Omni was not included, as it is only V.1.0 compliant. Fig. 5 contains one plot per compiler, including curves for most synchronization microbenchmarks. The results confirmed what we expected to see: increasing the number of threads in the second level leads to increased overheads. Due to space limitations, we cannot comment on every aspect of the plots but we believe that they present the situation very vividly. It is enough to say that for some implementations things seems to get out of control, especially for `parallel` and `reduction`. By far, the most scalable behavior is exhibited by the OMPi+PSTHREADS setup, although in absolute numbers the Intel compiler is in many cases the fastest.

5   Conclusion

In this paper we presented an extension to the EPCC microbenchmark suite that allows the measurement of OpenMP construct overheads under nested parallelism. Using this extension we studied the behavior of various OpenMP compilation and runtime systems when forced into inner parallel regions. We discovered that many implementations have scalability problems when nested parallelism is enabled and the number of threads increases well beyond the number of available processors. This is most probably due to the kernel-level thread model the majority of the implementations use. The utilization of kernel threads introduces significant overheads in the runtime library. When the number of threads that compete for hardware recourses significantly exceeds the number of available processors, the system is overloaded and the parallelization overheads outweigh any performance benefits. Finally, it becomes quite difficult for the runtime system to decide the distribution of inner-level threads to specific processors in order to favor computation and data locality.

Although our study was limited to two nesting levels, it became clear that studying deeper levels would only reveal worse behavior. It is evident that there are several design issues and performance limitations related to nested parallelism support that implementations have to address in an efficient way. In the near future we plan to expand the microbenchmark suite appropriately so as to be able to study the overheads at any arbitrary nesting level.

Acknowledgement. This work was co-funded by the European Union in the framework of the project "Support of Computer Science Studies in the University of Ioannina" of the "Operational Program for Education and Initial Vocational Training" of the 3rd Community Support Framework of the Hellenic Ministry of Education, funded by national sources and by the European Social Fund (ESF).

References

1. OpenMP Architecture Review Board: OpenMP C and C++ Application Program Interface, Version 2.5 (May 2005)
2. Bull, J.M.: Measuring Synchronization and Scheduling Overheads in OpenMP. In: Proc. of the 1st EWOMP, European Workshop on OpenMP, Lund, Sweden (1999)
3. Bull, J.M., O'Neill, D.: A Microbenchmark Suite for OpenMP 2.0. In: Proc. of the 3rd EWOMP, European Workshop on OpenMP, Barcelona, Spain (2001)
4. Ayguade, E., Gonzalez, M., Martorell, X., Jost, G.: Employing Nested OpenMP for the Parallelization of Multi-zone Computational Fluid Dynamics Applications. In: Proc. of the 18th Int'l Parallel and Distributed Processing Symposium, Santa Fe, New Mexico, USA (2004)
5. Blikberg, R., Sorevik, T.: Nested Parallelism: Allocation of Processors to Tasks and OpenMP Implementation. In: Proc. of the 28th Int'l Conference on Parallel Processing (ICCP 1999), Fukushima, Japan (1999)
6. Tanaka, Y., Taura, K., Sato, M., Yonezawa, A.: Performance Evaluation of OpenMP Applications with Nested Parallelism. In: Dwarkadas, S. (ed.) LCR 2000. LNCS, vol. 1915, pp. 100–112. Springer, Heidelberg (2000)
7. an Mey, D., Sarholz, S., Terboven, C.: Nested Parallelization with OpenMP. International Journal of Parallel Programming 35(5), 459–476 (2007)
8. Tian, X., Hoeflinger, J.P., Haab, G., Chen, Y.-K., Girkar, M., Shah, S.: A compiler for exploiting nested parallelism in OpenMP programs. Parallel Computing 31, 960–983 (2005)
9. Sun Microsystems: Sun Studio 12: OpenMP API User's Guide, PN819-5270 (2007)
10. Novillo, D.: OpenMP and automatic parallelization in GCC. In: Proc. of the 2006 GCC Summit, Ottawa, Canada (2006)
11. Thibault, S., Broquedis, F., Goglin, B., Namyst, R., Wacrenier, P.-A.: An Efficient OpenMP Runtime System for Hierarchical Architectures. In: Proc. of the 3rd Int'l Workshop on OpenMP (IWOMP 2007), Beijing, China (2007)
12. Sato, M., Satoh, S., Kusano, K., Tanaka, Y.: Design of OpenMP Compiler for an SMP Cluster. In: Proc. of the 1st EWOMP, European Workshop on OpenMP, Lund, Sweden (1999)
13. Dimakopoulos, V.V., Leontiadis, E., Tzoumas, G.: A Portable C Compiler for OpenMP V.2.0. In: Proc. of the 5th EWOMP, European Workshop on OpenMP, Aachen, Germany (2003)
14. Hadjidoukas, P.E., Dimakopoulos, V.V.: Nested Parallelism in the OMPi OpenMP C Compiler. In: Proc. of the European Conference on Parallel Computing (EU-ROPAR 2007), Rennes, France (2007)

CLOMP: Accurately Characterizing OpenMP Application Overheads*

Greg Bronevetsky, John Gyllenhaal, and Bronis R. de Supinski

Computation Directorate
Lawrence Livermore National Laboratory
Livermore, CA 94551, USA
greg@bronevetsky.com, gyllen@llnl.gov, bronis@llnl.gov

Abstract. Despite its ease of use, OpenMP has failed to gain widespread use on large scale systems, largely due to its failure to deliver sufficient performance. Our experience indicates that the cost of initiating OpenMP regions is simply too high for the desired OpenMP usage scenario of many applications. In this paper, we introduce CLOMP, a new benchmark to characterize this aspect of OpenMP implementations accurately. CLOMP complements the existing EPCC benchmark suite to provide simple, easy to understand measurements of OpenMP overheads in the context of application usage scenarios. Our results for several OpenMP implementations demonstrate that CLOMP identifies the amount of work required to compensate for the overheads observed with EPCC. Further, we show that CLOMP also captures limitations for OpenMP parallelization on NUMA systems.

1 Introduction

OpenMP [11] is a simple method to incorporate shared memory parallelism into scientific applications. While OpenMP has grown in popularity, it has failed to achieve widespread usage in those applications despite the use of shared memory nodes as the building blocks of large scale resources on which they run. Many factors contribute to this apparent contradiction, most of which reflect the failure of OpenMP-based applications to realize the performance potential of the underlying architecture. First, the applications run on more than one node of these large scale resources and, thus, the applications use MPI [10]. While distributed shared memory OpenMP implemetantions [9] are an option, they fail to provide the same level of performance.

Application programmers still might have adopted a hybrid OpenMP/MPI style, using OpenMP for on-node parallelization. However, the performance achieved discourages that also. OpenMP programs often have higher Amdahl's fractions than with MPI for on-node parallelization. Optimization of OpenMP

---

* This work performed under the auspices of the U.S. Department of Energy by Lawrence Livermore National Laboratory under Contract DE-AC52-07NA27344. (UCRL-ABS-XXXXXX).

R. Eigenmann and B.R. de Supinski (Eds.): IWOMP 2008, LNCS 5004, pp. 13–25, 2008.

usage has proven difficult due to a lack of a standard OpenMP profiling interface and, more so, to a myriad of confusing and often conflicting environment settings that govern OpenMP performance. In addition, the lack of on-node parallelization within MPI implementations has often implied higher network bandwidths with multiple MPI tasks on a node. Perhaps the most important factor has been a mismatch between the amount of work in typical OpenMP regions of scientific applications and the overhead of starting those regions.

Multi-core systems will impact many factors that have restricted adoption of OpenMP. Future networking hardware will not support the messaging rates required to achieve reasonable performance with an MPI task per core. Also, greater benefit from on-node parallelization within MPI implementations will provide similar (or better) aggregate network bandwidth to hybrid OpenMP/MPI applications compared to using an MPI task per core. Further, shared caches will provide memory bandwidth benefits to threaded applications.

Since we expect OpenMP to gain popularity with future large scale systems, we must understand the impact of OpenMP overheads on realistic application regions. Accurately characterizing them will help motivate chip designers to provide hardware support to reduce them if necessary. In this paper, we present CLOMP, a new benchmark that complements the EPCC suite [13] to capture the impact of OpenMP overheads (the CLOMP benchmark has no relationship to Intel's Cluster OpenMP). CLOMP is a simple benchmark that models realistic application code structure, and thus the associated limits on compiler optimization. We use CLOMP to model several application usage scenarios on a range of current shared memory systems. Our results demonstrate that OpenMP overheads limit performance substantially for large scale multiphysics applications and that NUMA effects can dramatically lower their performance even when they can compensate for those overheads.

2   Characteristics of Scientific Applications

CLOMP provides a single easy-to-use benchmark that captures the shared memory parallelization characteristics of a wide range of scientific applications. We focused on applications in use at Lawrence Livermore National Laboratory (LLNL), which are representative of large scale applications. We categorize LLNL applications as multiphysics applications or as science applications that focus on a particular physics domain. We need a simple easy-to-use benchmark that accurately characterizes the performance that a system and its OpenMP implementation will deliver to the full range of these applications.

Multiphysics applications [4,16,5,14] generally have large, complex code bases with multiple code regions that contribute significantly to their total run time. These routines occur in disparate application code sections as well as third party libraries, such as linear solvers [1,6]. While the latter may include large loops that are relatively amenable to OpenMP parallelization, the application code often has many relatively small but parallelizable loops with dependencies between the loops that inhibit loop fusion to increase the loop sizes. Further, the loops frequently occur in disparate function calls related to different physics packages, making consolidation even more difficult. Many multiphysics applications use unstructured grids, which imply significant pointer chasing to retrieve the actual data. Code restructuring to overcome these challenges is difficult: not only are these applications typically very large (a million lines of code or more) but the exact routines and the order in which they are executed depends on the input. However, the individual loops have no internal dependencies and would appear to be good candidates for OpenMP parallelization.

Science applications typically have fewer lines of code and less diverse execution profiles. While many still use high performance numerical libraries such as ScaLAPACK [2], a single routine often contains the primary computational kernel. Loop sizes available for OpenMP parallelization vary widely, from dense large matrix operations to very short loops. LLNL science applications include first principles molecular dynamics codes [8], traditional molecular dynamics codes [7,12,15] and ParaDiS, a dislocation dynamics application [3].

The loop sizes available for OpenMP parallelization depend on the application and the input problem. Currently, many HPC applications either use weak scaling or increase the problem resolution, both of which imply the loop sizes do not vary substantially as the total number of processors increases. However, we anticipate systems with millions of processor cores in the near future, which will make strong scaling attractive. Further, the amount of memory per core will decrease substantially. Both of these factors will lead to smaller OpenMP loops. Thus, while we need an OpenMP benchmark that characterizes the range of applications, capturing the impact of decreasing loop sizes is especially important.

3   The CLOMP Benchmark Implementation

CLOMP is structured like a multiphysics application. Its state mimics an unstructured mesh with a set of partitions, each divided into a linked list of zones, as Figure 1 shows. The linked lists limit optimizations but we allocate the zones contiguously so CLOMP can benefit from prefetching. The amount of memory allocated per zone can be adjusted to model different pressures on the memory system; however, computation is limited to the first 32 bytes of each zone. We kept the per-zone working set constant because many applications only touch a subset of a zone's data on each pass, including our target applications. Although the actual size varies from application to application, keeping it at 32 bytes makes it easier to explore the interactions between the CPU and the memory subsystem.

CLOMP repeatedly executes the loop shown in Figure 2. calc_deposit() represents a synchronization point, such as an MPI call or a computation that depends on the state of all partitions. The subsequent loop contains numPartitions independent iterations. Each iteration traverses a partition's linked list of zones, depositing a fraction of a substance into each zone. We tune the amount of computation per zone by repeating the inner loop flopScale times.

CLOMP models several possible loop parallelization methods, outlined in Figure 3. The first applies a combined parallel for construct to the outer loop, using either a `static` or a `dynamic` schedule. We call these configurations `for-static` and `for-dynamic`. The second method, called `manual`, represents parallelization that the programmer can perform manually to reduce the Amdahl's fraction. We enclose all instances of CLOMP's outer loop in a `parallel` construct and partition each work loop among threads explicitly. To ensure correct execution, we follow the work loop by a `barrier` and enclose the `calc_deposit` in a `single` construct. The last configuration, called `best-case` represents the optimistic scenario in which all OpenMP synchronization is instantaneous. It is identical to the `manual` version, except that the `barrier` and `single` are removed. Although this configuration would not produce correct answers, it provides an upper bound for the performance improvements possible for the other configurations.

While similar to the schedule benchmark in EPCC that measures the overhead of the loop construct with different schedule kinds, CLOMP emulates application scenarios through several parameters in order to characterize the impact of that overhead. The `numPartitions` parameter determines the number of independent pieces of work in each outer loop while the `numZonesPerPart` and the `flopScale` parameters determine the amount of work in each partition. While our results in Section 4.2 fix `numPartitions` to 64, we can vary it as appropriate for the application being modeled. The EPCC test fixes the corresponding factor at 128 per thread and requires source code modification to vary it; which prevents direct investigation of speed ups for a loop with a fixed total amount of work. The EPCC test also fixes the amount of work per iteration to approximately 100 cycles; our results show that this parameter directly impacts the speed up achieved. CLOMP could mimic the EPCC schedule benchmark through proper parameter settings but those would not correspond to any application scenarios likely to benefit from OpenMP parallelization.

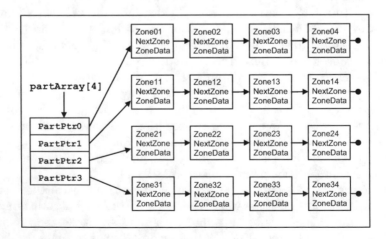

Fig. 1. CLOMP data structures

```
deposit = calc_deposit();
for(part = 0; part < numPartitions; part++) {
  for(zone = partArray[part]->firstZone; zone != NULL; zone = zone->nextZone) {
    for(scale_count = 0; scale_count < flopScale; scale_count++) {
      deposit = remaining_deposit * deposit_ratio;
      zone->value += deposit;
      remaining_deposit -= deposit; } } }
```

Fig. 2. CLOMP source code

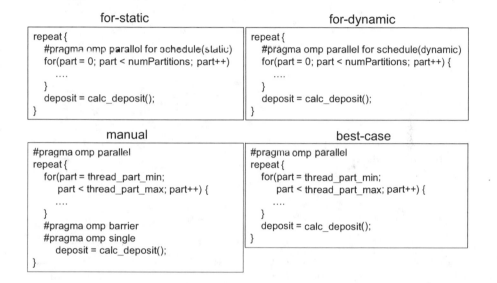

Fig. 3. Variants of CLOMP

Our results in Section 4.2 demonstrate that we must measure the impact of memory issues as well as the schedule overheads alone to capture the effectiveness of an OpenMP implementation for many realistic application loops. We control CLOMP's memory footprint through the `zoneSize` parameter that specifies the amount of memory allocated per zone. In addition, the `allocThreads` parameter determines whether each thread allocates its own partitions or if the master thread allocates all of the partitions. As is well known, the earlier strategy works better on NUMA systems that employ a first touch policy to place pages.

4   Experimental Results

In this section, we demonstrate that CLOMP provides the context of application OpenMP usage for results obtained with the EPCC microbenchmarks [13] through results on three different shared memory nodes. The LLNL Atlas system has dual core, quad socket (8-way) 2.4GHz Opteron, 16GB main memory nodes. Each core has 64KB L1 instruction and data caches and a 1MB L2 cache;

each dual core chip has a direct connection to 4GB of local memory with Hyper-
Transport connections to the memories of the other chips. The LLNL Thunder
system has 4-way 1.4GHz Itanium2, 4GB main memory nodes. Each single core
chip has 16KB instruction and data caches, a 256KB L2 cache and a 4MB L3
cache. All four processors on a node share access to main memory through four
memory hubs. Our experiments on Thunder and Atlas use the Intel compiler
version 9.1, including its OpenMP run time library support. The LLNL uP system has dual core, quad socket (8-way) 1.9GHz Power5, 32 GB main memory
nodes. Each core has private 64KB instruction and 32KB data caches while a
1.9MB L2 cache and a 36MB L3 cache are shared between the two cores on each
chip. Each dual core chip has a direct connection to 8GB of local memory with
connections through the other chips to their memories. Our experiments on uP
use the IBM xlc compiler version 7.0, including its OpenMP run time library
support.

All experiments on all platforms use the -O3 optimization level. Thread affinity
were used to force each thread to use a different core but the threads were
not bound, meaning that they could move if needed by the Operating System.
We relied on the kernel's memory affinity algorithm to keep memory close to
the threads that allocated it but the exact details of the algorithms used are
unknown.

4.1   OpenMP Overheads Measured with EPCC

We measured the overheads of OpenMP constructs on our target platforms with
the EPCC microbenchmark suite. Figure 4 presents the results of the synchronization microbenchmark and Figure 5 show the scheduling microbenchmark.
All figures list OpenMP constructs on the x-axis and their average overhead
from ten runs in processor cycles on the y-axis. The synchronization benchmark
data is plotted on a linear y-axis and the scheduling data uses a logarithmic axis.

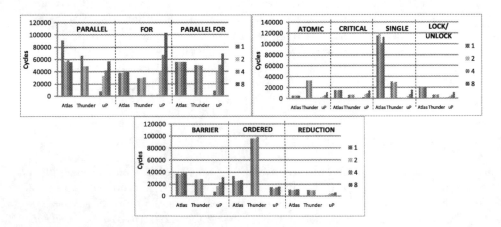

Fig. 4. EPCC Synchronization Results

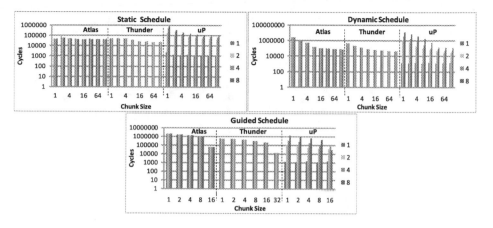

Fig. 5. EPCC Scheduling Results

The synchronization microbenchmark data shows several interesting effects. First, while the overhead of synchronization constructs with Intel OpenMP vary little with the number of threads, the overhead of the constructs with IBM OpenMP rises dramatically as the number of threads increases. However, despite its poor scaling, IBM OpenMP is less expensive for most OpenMP constructs The exceptions are the `atomic` and `critical` and parallel loop constructs, which have higher overhead with IBM OpenMP on larger thread counts. Overall, most synchronization overheads are on the order of tens of thousands of cycles. In particular, a `barrier` costs between 27,000 and 38,000 cycles with Intel OpenMP and from 7,000 to 31,000 with IBM OpenMP. The overhead of a loop construct is 28,000-40,000 cycles with Intel OpenMP and ranges from 1,400 to 100,000 cycles with IBM OpenMP. The overhead of a combined parallel loop construct is typically a little larger than the maximum overhead of the separate constructs.

The overhead of different schedule kinds varies between our platforms also, as shown in Figures 5 (the y-axis is logarithmic). The overhead of the loop construct changes little as the number of threads increases with our two Intel OpenMP platforms for a fixed schedule kind and associated chunk size. Further, static scheduling overhead is similar for all chunk sizes. In contrast, dynamic scheduling overhead drops off exponentially with increasing chunk size while guided scheduling overhead falls linearly. The reduced overheads reflect that the dynamic and guided mechanisms impose a cost every time they are invoked. Since larger chunks imply fewer invocations of the chunk assignment mechanism, they impose a smaller overhead. This drop-off is less pronounced for guided scheduling because it uses smaller chunks at the end of the allocation process, while dynamic scheduling uses similar chunk sizes throughout. Nonetheless, dynamic and guided scheduling overheads are consistently higher than static scheduling overhead on the Intel OpenMP platforms, ranging from twice as high with a chunk size of 128 to a factor of ten higher on Thunder and 50 on Atlas with a chunk size of one. On Thunder, guided scheduling overhead with a chunk size of 32 is 1.8x lower than the static scheduling overhead; the reason for this is unclear. The overheads of different schedule kinds with IBM OpenMP rise superlinearly with the number of threads. However, IBM OpenMP overheads exhibit the same patterns with respect to chunk size patterns as seen with Intel OpenMP except that static scheduling shows even steeper overhead drops than dynamic and guided scheduling with increasing the chunk size. In addition, static scheduling overhead is not much lower than the other schedule kinds with the same chunk size and is sometimes larger.

The EPCC results capture the relative cost of different schedule kinds on our platforms. When compared to Intel OpenMP, IBM OpenMP is always cheaper with dynamic and guided and one thread and is usually cheaper with two. In all other cases, IBM OpenMP is more expensive as its poor scalability overtakes its good sequential performance. The results demonstrate that users should use static scheduling with Intel OpenMP unless their loop bodies have very significant load imbalances while, with IBM OpenMP, the more flexible schedule kinds are more likely to prove worthwhile. However, these low level EPCC results do not include sufficient information to determine if an application can compensate for the overheads. While it helps to convert the overheads to cycles from the microseconds that the test suite reports, we still need measures that capture the effect of these overheads for realistic application scenarios.

4.2   Capturing the Impact of OpenMP Overheads with CLOMP

We model application scenarios through CLOMP parameter settings. All results presented here set `numPartitions` to 64 and `flopScale` to 1. CLOMP's default parameters, including `numZonesPerPart` equal to 100, model the relatively small loop sizes of many multiphysics application. The defaults use the minimum zone size of 32 bytes, which provides the most opportunity for prefetching and limits memory system pressure, and have the master thread allocate all memory similarly to the usual default in most applications.

The untuned results, shown in Figure 6, use the default run time environment variable settings, which is the most likely choice of application programmers. With these settings the (unrealistic) `best-case` configuration scales well up to 8 threads, which shows that good performance for the loop sizes common to multiphysics applications are possible. However, the realistic configurations all scale poorly, even causing increased run times in many cases.

The tuned results, shown in Figure 7, reflect the impact of changing environment settings so idle threads spin instead of sleep on uP and so idle threads spin much longer (`KMP_BLOCKTIME=100000`) before they sleep on Atlas and Thunder (we used these settings for the EPCC results presented in Section 4.1). These settings, which are appropriate for nodes dedicated to a single user, result in improved scaling for the `manual` and `for-static` scale configurations on both uP and Atlas. However, the actual speed ups, no more than 3.9, are still disappointing in light of the potential demonstrated by the `best-case` configuration. Further, the `for-dynamic` configuration still does not have sufficient work to compensate for the high overhead of the dynamic schedule kind. In fact, the "tuned"

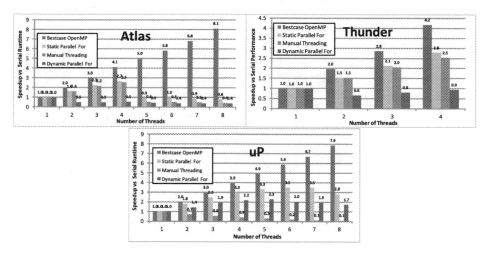

Fig. 6. CLOMP Untuned Default Scenario

environment settings actually caused a slowdown for `for-dynamic` on uP and they did not improve performance on Thunder. These results highlight the complexity of choosing the best OpenMP configuration, a task for which CLOMP results provide guidance. For our subsequent experiments we consistently used the modified OpenMP flags because the best performance of `for-dynamic` is much lower than the best performance of `manual` and `for-static`.

We examined the effects of memory bandwidth on the performance of parallel loops by increasing the number of zones per partition by a factor of 10 (1,000 zones per partition), which corresponds to some multiphysics application runs as well as some science codes. The results for this scenario, shown in Figure 8, exhibit outstanding scaling since the single core's memory bandwidth dominates performance of the sequential run. In fact, we observe superlinear speedups with `manual` and `for-static` on uP (e.g., 8.7x on with 8 threads) and on Atlas (peaking at 36x on 7 threads). The dramatic improvement on Atlas arises from the system's NUMA architecture, in which the penalty for accessing remote memory via Hyper-Transport is relatively very high. Since the problem fits in cache with more threads, the performance benefit is significant. The cache effects are far smaller on uP and Thunder since these systems provide uniform memory access, with uP's slightly super-linear speedups attributable to its much larger cache. In all cases, these configurations are very close to the theoretical maximum of `best-case` while the `for-dynamic` configuration results continue to disappoint.

For application scenarios with even larger memory footprints, corresponding to science codes based on dense linear algebra routines, we no longer observe superlinear speedups since they no longer fit into cache. However, while we observe consistently good scaling on the uniform memory access systems, these scenarios provide insight into NUMA performance issues. Figure 9 shows results on Atlas for scenarios in which we increase the number of zones per partition over the default scenario 100x (10,000 zones per partition) and 1,000x (100,000 zones

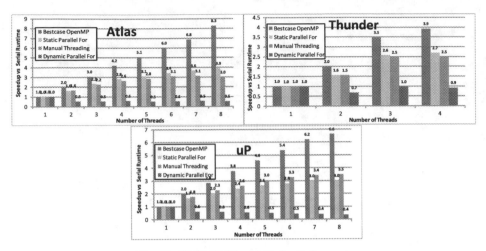

Fig. 7. CLOMP Tuned Default Scenario

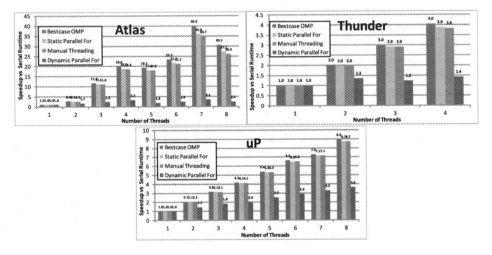

Fig. 8. CLOMP 10X Memory Scenario

per partition). Here, we compare the two strategies for allocating application state: serial, where the master thread allocates all memory; and threaded, where each thread allocates its own memory. For each allocation strategy we show the speedup of the highest-performing realistic configuration. In both scenarios the two allocation strategies result in dramatically different performance, with the threaded allocation achieving near-linear speedup, while the serial allocation shows little improvement at all scales, similarly to previous observations on other NUMA systems. While application programmers generally will make the necessary coding changes to achieve these performance gains, the gains are not consistent: we still observed significant performance variation in our runs,

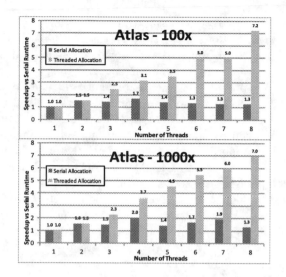

Fig. 9. CLOMP 100X and 1000X Memory Scenarios

with speed ups as low as 4 with eight threads. Examination of /proc data indicates that the threaded allocation does not guarantee the strict use of local memory. We are investigating using the numactl command in the NUMA library to provide more consistent performance.

By providing a best-case performance estimate, CLOMP puts the actual performance numbers in context of OpenMP overheads, cache effects, and NUMA effects. The best-case configuration is significantly different from the EPCC schedule test and represents a contribution of our work. For example, in Figure 8, the 27.1 speedup for 8 threads on Atlas is great but an even higher speedup of 30.5 was possible if the OpenMP overheads were lower. Similarly, the low best-case serial Allocation performance corresponding to the results in Figure 9 shows that OpenMP overhead is not the problem, NUMA effects are.

5   Conclusion and Future Work

Despite the popularity of shared memory systems and OpenMP's ease of use, overheads in OpenMP implementations and shared memory hardware have limited potential performance gains, thus discouraging the use of OpenMP. This paper presents CLOMP, a new OpenMP benchmark that models the behavior of scientific applications that have an overall sequential structure but contain many loops with independent iterations. CLOMP can be parameterized to represent a variety of applications, allowing application programmers to evaluate possible parallelization strategies with minimal effort and OpenMP implementors to identify overheads that can have the largest impact on real applications. Our results on three shared memory platforms demonstrate that CLOMP extends EPCC to capture the application scenarios necessary to characterize the impact of the overheads measured by EPCC. CLOMP guides selection of run time environment settings and can identify the impact of architectural features such as memory bandwidth and a NUMA architecture on application performance. The resulting insights can be very useful to application programmers in choosing the parallelization strategy and hardware that will provide the best performance for their application.

CLOMP is focused on single-node application performance while most scientific applications execute on multiple nodes, using MPI for inter-node communication. Our experience indicates that environment settings appropriate for single node OpenMP applications are often detrimental to MPI performance. Thus, our future work will extend CLOMP to include MPI communication so we can analyze the performance trade-offs between OpenMP and MPI.

Overall, our results should not be seen as critiquing the OpenMP implementations that were used in our experiments. While we noted differences between them, the most significant issues arose from differences in the underlying architecture. Ultimately, CLOMP would provide its greatest value if it could guide architectural refinements that reduce the overheads of dispatching threads for OpenMP regions. For this reason, we are including CLOMP in the benchmark suite of LLNL's Sequoia procurement.

References

1. Balay, S., Gropp, W.D., McInnes, L.C., Smith, B.F.: Efficient management of parallelism in object oriented numerical software libraries. In: Arge, E., Bruaset, A.M., Langtangen, H.P. (eds.) Modern Software Tools in Scientific Computing, pp. 163–202. Birkhäuser Press (1997)
2. Blackford, L., Choi, J., Cleary, A., Azevedo, E., Demmel, J., Dhillon, I., Dongarra, J., Hammerling, S., Henry, G., Petite, A., Stanley, K., Walker, D., Whaley, R.: ScaLAPACK Users. In: SIAM, Philadelphia (1997)
3. Bulatov, V., Cai, W., Fier, J., Hiratani, M., Hommes, G., Pierce, T., Tang, M., Rhee, M., Yates, K.R., Arsenlis, T.: Scalable Line Dynamics in ParaDiS. In: Proceedings of IEEE/ACM Supercomputing 2004 (November 2004)
4. Collins, W.D., Bitz, C.M., Blackmon, M.L., Bonan, G.B., Bretherton, C.S., Carton, J.A., Chang, P., Doney, S.C., Hack, J.J., Henderson, T.B., Kiehl, J.T., Large, W.G., McKenna, D.S., Santer, B.D., Smith, R.D.: The community climate system model version 3. Journal of Climate 19(1), 2122–2143 (2006)
5. de St. Germain, J.D., McCorquodale, J., Parker, S.G., Johnson, C.R.: A Component-based Architecture for Parallel Multi-Physics PDE Simulation. In: International Symposium on High Performance and Distributed Computing (2000)
6. Falgout, R., Jones, J., Yang, U.: The Design and Implementation of HYPRE, a Library of Parallel High Performance Preconditioners. Numerical Solution of Partial Differential Equations on Parallel Computers. Springer, Heidelberg (to appear)
7. Germann, T., Kadau, K., Lomdahl, P.: 25 Tflop/s Multibillion-Atom Molecular Dynamics Simulations and Visualization/Analysis on BlueGene/L. In: Proceedings of IEEE/ACM Supercomputing 2005 (November 2005)
8. Gygi, F., Draeger, E., de Supinski, B.R., Yates, R.K., Franchetti, F., Kral, S., Lorenz, J., Überhuber, C.W., Gunnels, J.A., Sexton, J.C.: Large-Scale First-Principles Molecular Dynamics simulations on the BlueGene/L Platform using the Qbox code. In: Proceedings of IEEE/ACM Supercomputing 2005 (November 2005)

9. Hoeflinger, J., de Supinski, B.R.: The openmp memory model. In: International Workshop on OpenMP (IWOMP) (2005)
10. Message Passing Interface Forum. Mpi: A message-passing interface standard. International Journal of Supercomputer Applications 8(3/4), 165–414 (1994)
11. OpenMP Architecture Review Board. OpenMP application program interface, version 2.5
12. Phillips, J.C., Zheng, G., Kumar, S., Kale, L.V.: NAMD: Biomolecular Simulation on Thousands of Processors. In: Proceedings of IEEE/ACM Supercomputing 2002 (November 2002)
13. Reid, F.J.L., Bull, J.M.: Openmp microbenchmarks version 2.0. In: European Workshop on OpenMP (EWOMP) (2004)
14. Rosner, R., Calder, A., Dursi, J., Fryxell, B., Lamb, D.Q., Niemeyer, J.C., Olson, K., Ricker, P., Timmes, F.X., Truran, J.W., Tufo, H., Young, Y.-N., Zingale, M., Lusk, E., Stevens, R.: Flash code: Studying astrophysical thermonuclear flashes. Journal on Computing in Science and Engineering 2(2) (2000)
15. Streitz, F., Glosli, J., Patel, M., Chan, B., Yates, R., de Supinski, B., Sexton, J., Gunnels, J.: 100+ TFlop Solidification Simulations on BlueGene/L. In: Proceedings of IEEE/ACM Supercomputing 2005 (November 2005)
16. White, B.S., McKee, S.A., de Supinski, B.R., Miller, B., Quinlan, D., Schulz, M.: Improving the Computational Intensity of Unstructured Mesh Applications. In: Proceedings of the 19th ACM International Conference on Supercomputing (June 2005)

Detection of Violations to the MPI Standard in Hybrid OpenMP/MPI Applications

Tobias Hilbrich[1], Matthias S. Müller[1], and Bettina Krammer[2]

[1] TU Dresden, Center for Information Services and
High Performance Computing (ZIH), 01062 Dresden, Germany
[2] High Performance Computing Center (HLRS), Universität Stuttgart,
Nobelstrasse 19, 70569 Stuttgart, Germany

Abstract. The MPI standard allows the usage of multiple threads per process. The main idea was that an MPI call executed at one thread should not block other threads. In the MPI-2 standard this was refined by introducing the so called *level of thread support* which describes how threads may interact with MPI. The multi-threaded usage is restricted by several rules stated in the MPI standard. In this paper we describe the work on an MPI checker called MARMOT[1] to enhance its capabilities towards a verification that ensures that these rules are not violated. A first implementation is capable of detecting violations if they actually occur in a run made with MARMOT. As most of these violations occur due to missing thread synchronization it is likely that they don't appear in every run of the application. To detect whether there is a run that violates one of the MPI restrictions it is necessary to analyze the OpenMP usage. Thus we introduced artificial data races that only occur if the application violates one of the MPI rules. By this design all tools capable of detecting data races can also detect violations to some of the MPI rules. To confirm this idea we used the Intel® Thread Checker.

1 Introduction

1.1 Hybrid MPI Applications

The MPI standard states that MPI calls should only block the calling thread. This was refined in the MPI-2 standard by introducing the so called *level of thread support* (*thread level*). Each MPI implementation supports the lowest level and may support a higher one. The levels are:

MPI_THREAD_SINGLE:	only one thread exists
MPI_THREAD_FUNNELED:	multiple threads may exist but only the main[1] thread performs MPI calls
MPI_THREAD_SERIALIZED:	multiple threads exist and each thread may perform MPI calls as long as no other thread is calling MPI
MPI_THREAD_MULTIPLE:	multiple threads may call MPI simultaneously

---

[1] The thread that initialized MPI.

R. Eigenmann and B.R. de Supinski (Eds.): IWOMP 2008, LNCS 5004, pp. 26–35, 2008.

The application specifies a desired *thread level* and passes it to the MPI implementation, which might return a lower level. The level returned is referred to as provided *thread level* and must not be violated.

1.2   Restrictions for Hybrid MPI Applications

Besides the definition of the *thread level* there are further restrictions mentioned in the MPI standard. Here we present the restrictions that are currently checked by our MPI checker. This list is incomplete, e.g. most MPI-2 calls are currently unsupported. The currently checked restrictions are:

(I) *The call to MPI_FINALIZE should occur on the same thread that initialized MPI. We call this thread the main thread. The call should occur only after all the process threads have completed their MPI calls, and have no pending communications or I/O operations.* [2, page 194 lines 24-29]

(II) *A program where two threads block, waiting on the same request, is erroneous. Similarly, the same request cannot appear in the array of requests of two concurrent MPI_WAIT{ANY|SOME|ALL} calls. In MPI, a request can only be completed once. Any combination of wait or test which violates this rule is erroneous.* [2, page 194 lines 32-36]

(III) *A receive call that uses source and tag values returned by a preceding call to MPI_PROBE or MPI_IPROBE will receive the message matched by the probe call only if there was no other matching receive after the probe and before that receive. In a multithreaded environment, it is up to the user to enforce this condition using suitable mutual exclusion logic. This can be enforced by making sure that each communicator is used by only one thread on each process.* [2, page 194f line 46ff]

(IV) *Finally, in multithreaded implementations, one can have more than one, concurrently executing, collective communication call at a process. In these situations, it is the usere's responsibility to ensure that the same communicator is not used concurrently by two different collective communication calls at the same process.* [3, page 130 lines 37-41]

We will refer to these restrictions by using their respective number. The additional restriction that the provided *thread level* may not be violated will be referred to as (V). To our knowledge none of the available MPI Checkers does check any of these restrictions. The MPI checkers we looked at are: Umpire[4], MPI-Check[5] and Intel® Message Checker[6].

2   Example

For restriction (II) we give an example violation in Table 1. Violations to this restriction are only possible if the *thread level* MPI_THREAD_MULTIPLE is available. This applies to most of the restrictions. In the example, process 1 has two threads that are simultaneously performing an MPI_Wait and an MPI_Test

Table 1. Violation to restriction (II), assuming that *request* is a shared variable

Process 1	
Thread 1	Thread 2
MPI_Isend(msg,&request)	
#pragma omp barrier	#pragma omp barrier
MPI_Wait(request)	MPI_Test(request)

call. Both calls are using the same request. This violates restriction (II). As one can see if the MPI_Test call is performed before the MPI_Wait call the error won't occur whereas simultaneous execution yields an error.

3   Constraints for the Restrictions

In order to create checks for restrictions (I) to (V) we want to define constraints that implicate whether one of the restrictions is violated. These constraints can be used to implement checks and to verify whether the implemented checks can detect every instance of the problem. If a run of an application matches any of these constraints this implies that there is a violation to the respective restriction. The constraints are:

violation to (I):
    If and only if one of the following holds:
    (A) for a thread calling MPI_Finalize holds: "MPI_Is_thread_main() == False"
    (B) one thread is performing an MPI_Finalize call while another thread is also calling MPI
    (C) an MPI call is issued after the call to MPI_Finalize is finished

violation to (II):
    If and only if a thread performs a *Wait*[2] or *Test*[3] call using Request X while another thread is also using Request X in a *Wait* or *Test* call.

violation to (III):
    If and only if one of the following holds:
    (A) one thread is performing a *Probe*[4] call that will return the value X as *source* and the value Y as *tag* and another thread is performing a *Recieve*[5] call with *source* and *tag* values that match X and Y at the same time
    (B) a thread is performing a *Receive* call with source and tag values returned by a *Probe* call of another thread that did not receive that message yet

---

[2] An MPI_Wait{ANY|ALL|SOME} call.
[3] An MPI_Test{ANY|ALL|SOME} call.
[4] An MPI_Iprobe or an MPI_Probe call.
[5] An MPI_Recv, MPI_IRecv, MPI_Sendrecv, MPI_Sendrecv_replace or (MPI_Start) call.

(C) one thread is receiving a message with *source* and *tag* values returned
by a *Probe* it called and another thread is receiving a message with
matching *source* and *tag* values at the same time (assumed that the
probed message was not yet received)

violation to (IV):

If and only if a thread performs a collective call using communicator X
while another thread is within a collective call using communicator X

violation to (V):

If and only if one of the following holds:

(A) provided is MPI_THREAD_SINGLE, omp_in_parallel() returns true
and omp_get_num_threads() returns a value greater than 1 while the
application is in an MPI call

(B) provided is MPI_THREAD_FUNNELED and MPI_Is_thread_main() returns false while application is in an MPI call

(C) provided is MPI_THREAD_SERIALIZED and two threads are calling
MPI simultaneously

One might have compressed constraints (A)-(C) of restriction (III) into one constraint which detects whether a *Receive* call is issued between a *Probe* and a *Receive* call of another thread (all using the same *source* and *tag* values). But we will
see later that this decomposition is useful when creating artificial data races.

4   Applying an MPI Checker to a Hybrid Application

4.1   Instrumenting the Application

Runtime MPI checkers have to monitor all performed MPI calls and check their
parameters and results. This is usually achieved by intercepting the MPI calls and
executing additional code before and after execution of the MPI call. For this purpose the MPI standard specifies the so called Profiling Interface. For each MPI
call "MPI_X" there is a second function "PMPI_X" that executes the same call.
Thus it is possible to create wrappers that catch each MPI call and execute additional code before and after calling the appropriate PMPI call. This is illustrated
in Fig. 1. We will refer to the code executed before the PMPI Call as the *pre execution code* and to the code after the PMPI call as the *post execution code*.

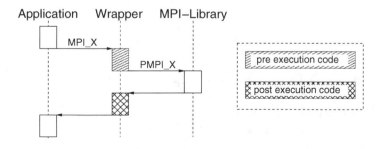

Fig. 1. Execution of an MPI call when using a wrapper

4.2   Propositions on Correctnes

Using *pre* and *post execution code* in an environment in which multiple threads are calling MPI simultaneously might change the order in which MPI calls are issued by the threads. Imagine a *pre execution code* that synchronizes the threads such that only one thread is calling MPI at a time. That would prohibit violations to most of the above mentioned restrictions. It is desirable that the *pre* and *post execution code* is designed such that it does not change the semantics of the application. This especially requires that if there is a run of the application in which the commands are executed in a certain order between the threads then there is a run with the wrapper attached in which the commands of the application are executed in the same order. We propose several rules for the *pre* and *post execution code* that should enforce preservation of semantics:

- the code must not enforce serialization or a certain ordering of the MPI calls made on the threads of each process
- the code must not enforce an ordering of the MPI calls of different processes
- execution may only take a finite time
- parameters passed to the MPI call must not be modified
- no data that is used by the application or the MPI library must be changed
- no input files used by the application or the MPI library must be changed

We assume that as long as the code within the wrappers does not violate any of these restrictions all errors that appear in the application may still appear when the wrapper is used. However, the probabilities of errors to occur during a run with or without wrapper might differ.

4.3   Synchronization

Our MPI checker uses process local data to store whether certain MPI events have occurred and to track the usage of MPI resources. In a multi threaded environment it is necessary to protect these data against unsynchronized access. To do so we introduced a mutual exclusion mechanism that is executed in the *pre* and *post execution code*. The scheme used is shown in Fig. 2. Before any process global data is used in the *pre execution code* the synchronization is started with a call to "enterMARMOT". It is stopped at the end of the *pre execution code* by calling "enterPMPI". The same scheme is used for the *post execution code*. To implement the mutual exclusion we use OpenMP locks. This synchronization should not violate any of the above propositions thus all MPI errors of the Application can still appear when our MPI wrappers are used.

5   First Implementation of Checks

As a first approach we implemented checks that detect violations if they actually occur in a run with MARMOT. Therefore we added *pre* and *post execution code*. Note that some of the code might be executed before the mutual exclusion starts and might thus require additional synchronization. Implementation

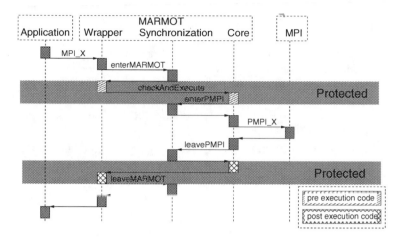

Fig. 2. Scheme of the applied synchronization within MARMOT

of most checks is straightforward and requires only a few lines of code. To implement checks for the constraints of restrictions (II) and (IV) we register used requests and communicators. We illustrate the used technique for the case of the communicators. In the *pre execution code* of all collective calls we check whether the given communicator is already registered as used. If that's the case we return an error, otherwise we register the communicator as used. In the *post execution code* we remove the communicator from the registration. In order to create a check for the constraints of restriction (III) we use a mechanism that registers *source* and *tag* values returned by a probe. These values are registered until the message is received. If the receiving thread was not the thread that probed for the message we issue a warning. With these checks we were able to detect violations to the restrictions in small test applications.

6   Detection with Artificial Data Races

6.1   General Design

The problem of the first detection is that it can only detect violations if they actually appear in a run made with MARMOT. It is desirable to also detect whether it is possible that a violation might ever appear. For the MPI calls of a hybrid application different runs may have a different execution order of the MPI calls and it might happen that certain MPI calls are issued by different threads. The execution order is important for constraint (B) and (C) of restriction (I), for all constraints of restrictions (II), (III), (IV) and for constraint (C) of restriction (V). For all other constraints it is only of interest which thread is calling MPI or how many threads are used. Note that for the constraints of restriction (III) it is also important which threads are performing the MPI calls.

Assume that each thread writes one and the same variable in the *pre execution code* of the Wrapper. Then there is a data race[7] on this variable if it is possible that two threads are issuing an MPI call simultaneously. This idea can be used to create artificial data races that only exist if a violation to one of the constraints is given. As there are tools that are capable to detect data races it is possible to determine whether a violation of one of the constraints occurs by detecting the associated data race. This technique can be applied to all the constraints that require an unsynchronized execution of MPI calls to happen. As detecting the violations relies on detecting a data race with a third party tool, results will only be as good as the tool applied.

Another aspect of this design is that it is necessary to create these races either before the synchronization starts or to stop and restart the synchronisation for the race.

6.2   Detection of *Thred Level* Violation

For constraint (C) of this restriction it is simply necessary to write a variable in the *pre execution code* of each MPI call. Note that most tools that detect data races have a fixed output format and will thus only issue an error that there is a data race on a certain variable. To influence this output we can only change the name of the variable that caused the data race. For this constraint we use the variable "App_Needs_MPI_THREAD_MULTIPLE".

6.3   Detection of Wrong Communicator Usage

To detect a violation to constraint (IV) it is necessary to detect whether it is possible that two collective calls use the same communicator simultaneously. Thus we have to design a data race that only occurs when this restriction is violated. We achieved this by mapping each communicator to an index. The index is used to write the corresponding entry of an array. In this way we only get a data race if two collective calls use the same communicator simultaneously. The *pre execution code* code used has the following structure (this code is executed before the synchronisazion starts):

```
beginCritical()
id = communicator2id(comm)
endCritical()
writeCommVarIndexed(id)
```

It is necessary to create a critical section here as mapping a communicator to an index requires the usage of an internal list that stores all the known communicators and their respective indizes. This list must only be used by one thread at a time to avoid unintended data races.

6.4   Detection of MPI_Probe Invalidation

To detect violations to this restriction it is necessary to create checks for the associated constraints (A),(B) and (C). To illustrate the need to check all three

Table 2. Different instances of violations to restriction (III)

Case 1	
Thread 1	Thread 2
MPI_Probe	MPI_Recv
MPI_Recv	

Case 2	
Thread 1	Thread 2
MPI_Probe	
omp barrier	omp barrier
MPI_Recv	MPI_Recv

Case 3	
Thread 1	Thread 2
MPI_Probe	MPI_Recv
omp barrier	omp barrier
MPI_Recv	

Case 4	
Thread 1	Thread 2
MPI_Probe	
omp barrier	omp barrier
	MPI_Recv
omp barrier	omp barrier
MPI_Recv	

constraints we present four different instances of this problem in table 2. Assume that all calls are using the same values for *source, tag* and *comm*. For each of these instances one of the three constraints is violated. Especially note that in all cases except the fourth either constraint (A) or constraint (C) is violated. Due to the synchronization present in the fourth case constraint (B) will be violated for all runs of this case. Thus the already implemented detection for constraint (B) is sufficient. So it is only necessary to create data races for constraints (A) and (C). Detecting violations to (A) is hard as the result of the *Probe* call might differ between runs. Thus it would be necessary to know all the possible return values of the *Probe* call. We avoided this as restriction (III) only yields a warning. Thus we only created data races that do not cover all instances of the problem and that might be hard to detect by a tool. Better data races could be created by using a more complicated scheme. The simple data races use the following *pre* and *post execution codes*:

```
Post execution code for all Probe calls:
conflict_index = getNextIndex()
registerNewProbe(returned_source,returned_tag,comm,conflict_index)
endSynchronization()
writeProbeVarIndexed(conflict_index)
beginSynchronization()

Pre execution code for all Receive calls:
FORALL probes B in registered probes {
    if (matchesProbe(B,my_source,my_tag,my_comm)) {
        endSynchronization()
        writeProbeVarIndexed(B->conflict_index)
        beginSynchronization()
        unregisterProbe(B)
    }
}
```

With this design there is a data race between two simultaneously executed Receive calls if one of them is invalidating a *Probe* call. There is also a data race if a *Receive* call is called in parallel to a *Probe* call that returned *source* and *tag* values matched by the receive. This race is hard to detect as not all runs will enter the code in the if statement.

6.5   Detection of Wrong Request Usage

A data race for the constraint of restriction (II) can be constructed as in the case of restriction (IV). Again it is necessary to map each request to an index in an array of variables. Some of the *Wait* and *Test* calls can use an array of Requests. In this case one has to map each request to an index and write the respective variable.

6.6   Detection of Erroneus MPI_Finalize

In order to create checks for the constraints of restriction (I) it is only necessary to create a data race for constraint (B). This is achieved by reading a variable for each MPI call and writing it in the *pre execution code* of MPI_Finalize. Thus if the application has the possibility to execute another MPI call in parallel to MPI_Finalize then there exists a data race.

7   Results with Intel® Thread Checker

The Intel® Thread Checker is a tool capable of detecting data races and thus should be able to detect the artificial races. We used small test codes that violate one of restrictions (I)-(V) to test this approach. In almost all the tests Thread Checker detected the data race. Only some violations to restriction (III) could not be detected. We want to present the gained output for a violation to restriction (IV). The output is shown in Table 3. As the data race is caused by code within our MPI checker the output points to source code of MARMOT. In order

Table 3. Thread Checker output for a violation to restriction (IV)

Intel(R) Thread Checker 3.0 command line instrumentation driver (23479) Copyright (c) 2006 Intel Corporation. All rights reserved.

ID	...	Severity Name	Count	...	Description	1st Access[Best]	...
					...		
2	...	Error	1	...	Memory write of this→App_Comm_In_Two_Collective_Calls[] at "mpo_TCheck_Races.cc":105 conflicts with a prior memory write of this→App_Comm_In_Two_Collective_Calls[] at "mpo_TCheck_Races.cc":105 (output dependence)	"mpo_TCheck_Races.cc":105	...
					...		

to find out where the error occurred in the application one has to use the stack trace feature of Thread Checker.

But usage of this tool also has disadvantes. OpenMP applications using source code instrumentation of Thread Checker are only executed on one thread. Thus for our MPI checker only one thread is visible which makes it impossible to directly detect violations to the restrictions. When using binary instrumentation of Thread Checker, execution uses multiple threads but the results of the tool are less precise. We could still detect the data races in this case but the results do not contain the variable names used within MARMOT which makes it hard to interpret the results.

8   Conclusion

We presented an approach to find bugs in hybrid OpenMP/MPI applications that violate restrictions in the MPI standard. A first implementation in the MPI checker MARMOT is able to detect violations that actually occur in a run. In order to detect race conditions and thus potential violations we introduced a technique with artifical data races. Construction of these races is only necessary for a subset of the constraints. The data races are detectable by appropriate shared memory tools. We demonstrated our approach with the Intel® Thread Checker.

References

1. Krammer, B., Bidmon, K., Müller, M.S., Resch, M.M.: MARMOT: An MPI Analysis and Checking Tool. In: Joubert, G.R., Nagel, W.E., Peters, F.J., Walter, W.V. (eds.) PARCO. Advances in Parallel Computing, vol. 13, pp. 493–500. Elsevier, Amsterdam (2003)
2. Message Passing Interface Forum: MPI-2: Extensions to the Message-Passing Interface (1997), http://www.mpi-forum.org/docs/mpi-20.ps
3. Message Passing Interface Forum: MPI: A Message-Passing Interface Standard (1995), http://www.mpi-forum.org/docs/mpi-10.ps
4. Vetter, J.S., de Supinski, B.R.: Dynamic software testing of MPI applications with umpire. In: Supercomputing 2000: Proceedings of the 2000 ACM/IEEE conference on Supercomputing (CDROM), Washington, DC, USA, p. 51. IEEE Computer Society, Los Alamitos (2000)
5. Luecke, G.R., Zou, Y., Coyle, J., Hoekstra, J., Kraeva, M.: Deadlock detection in MPI programs.. Concurrency and Computation: Practice and Experience 14(11), 911–932 (2002)
6. DeSouza, J., Kuhn, B., de Supinski, B.R., Samofalov, V., Zheltov, S., Bratanov, S.: Automated, scalable debugging of MPI programs with Intel® Message Checker. In: SE-HPCS 2005: Proceedings of the second international workshop on Software engineering for high performance computing system applications, pp. 78–82. ACM, New York (2005)
7. Netzer, R.H.B., Miller, B.P.: What are race conditions?: Some issues and formalizations. ACM Lett. Program. Lang. Syst. 1(1), 74–88 (1992)

Early Experiments with the
OpenMP/MPI Hybrid Programming Model

Ewing Lusk[1,*] and Anthony Chan[2]

[1] Mathematics and Computer Science Division
Argonne National Laboratory
[2] ASCI FLASH Center
University of Chicago

Abstract. The paper describes some very early experiments on new architectures that support the hybrid programming model. Our results are promising in that OpenMP threads interact with MPI as desired, allowing OpenMP-agnostic tools to be used. We explore three environments: a "typical" Linux cluster, a new large-scale machine from SiCortex, and the new IBM BG/P, which have quite different compilers and runtime systems for both OpenMP and MPI. We look at a few simple, diagnostic programs, and one "application-like" test program. We demonstrate the use of a tool that can examine the detailed sequence of events in a hybrid program and illustrate that a hybrid computation might not always proceed as expected.

1 Introduction

Combining shared-memory and distributed-memory programming models is an old idea [21]. One wants to exploit the strengths of both models: the efficiency, memory savings, and ease of programming of the shared-memory model and the scalability of the distributed-memory model. Until recently, the relevant models, languages, and libraries for shared-memory and distributed-memory architectures have evolved separately, with MPI [7] becoming the dominant approach for the distributed-memory, or message-passing, model, and OpenMP [9,15] emerging as the dominant "high-level" approach for shared memory with threads. We say "high-level" since it is higher level than the POSIX `pthread` specification [8]. We use quotation marks around the expression because OpenMP is not as high level as some other proposed languages that use models with a global view of data [1,19].

* This work was supported in part by the U.S. Department of Energy Contract #B523820 to the ASC/Alliance Center for Astrophysical Thermonuclear Flashes at the University of Chicago and in part by the Mathematical, Information, and Computational Sciences Division subprogram of the Office of Advanced Scientific Computing Research, Office of Science, U.S. Departent of Energy, under Contract DE-AC02-06CH11357.

R. Eigenmann and B.R. de Supinski (Eds.): IWOMP 2008, LNCS 5004, pp. 36–47, 2008.
© Springer-Verlag Berlin Heidelberg 2008

Recently, the hybrid model has begun to attract more attention, for at least two reasons. The first is that it is relatively easy to pick a language/library instantiation of the hybrid model: OpenMP plus MPI. While there may be other approaches, they remain research and development projects, whereas OpenMP compilers and MPI libraries are now solid commercial products, with implementations from multiple vendors. Moreover, we demonstrate in this paper that OpenMP and MPI implementations on significant platforms work together as they should (see Section 2.) The second reason is that scalable parallel computers now appear to encourage this model. The fastest machines now virtually all consist of multi-core nodes connected by a high speed network. The idea of using OpenMP threads to exploit the multiple cores per node (with one multithreaded process per node) while using MPI to communicate among the nodes appears obvious. Yet one can also use an "MPI everywhere" approach on these architectures, and the data on which approach is better is confusing and inconclusive. It appears to be heavily dependent on the hardware, the OpenMP and MPI implementations, and above all on the application and the skill of the application writer.

We do not intend to settle here the question of whether the hybrid approach is good or bad. Postings to assorted discussion lists claim to have proven both positions. Instead we describe three interesting environments in which this question can be studied, and present some preliminary experiments.

Considerable work has gone into studying the hybrid model. Some examples can be found in [12,18,20]. What is new in this paper is 1) a discussion of the relationship between the MPI standard and the OpenMP standard, 2) presentation of early results on two brand-new machines that support the hybrid model in an efficient way, and 3) depiction of a particular performance visualization tool looking at hybrid codes.

The rest of the paper is organized as follows. In Section 2 we describe aspects of MPI and OpenMP that pertain to their use with each other. In Section 3 we describe a performance visualization tool and recent enhancements to it that enable it to be used to study the behavior of hybrid programs. Section 4 describes the three environments that we used for our experiments, and early results from our experiments are in Section 5. We present some conclusions and plans for future work in Section 6.

2   What the Standards Say

Hybrid programming with two portable APIs would be impossible unless each made certain commitments to the other on how they would behave. In the case of OpenMP, the important commitment is that if a single thread is blocked by an operating system call (such as file or network I/O) then the remaining threads in that process will remain runnable. In our situation, this means that an MPI call that may block, such as `MPI_Recv` or `MPI_Wait`, will only block the calling thread and not the entire process. This is a significant commitment, since it involves the thread scheduler in the compiler's runtime system and interaction with the operating system.

The commitments made by the MPI standard are more complex. The MPI-2 standard defines four levels of thread safety. These are in the form of what commitments the application makes to the MPI implementation.

MPI_THREAD_SINGLE There is only one thread in the application.
MPI_THREAD_FUNNELED There is only one thread that makes MPI calls.
MPI_THREAD_SERIALIZED Multiple threads make MPI calls, but only one at a time.
MPI_THREAD_MULTIPLE Any thread may make MPI calls at any time.

An application can find out at run time which level is supported by the MPI library it is linked with by means of the MPI_Init_thread function call.

These levels correspond to the use of MPI in an OpenMP program in the following way:

MPI_THREAD_SINGLE There is no OpenMP multithreading in the program.
MPI_THREAD_FUNNELED All of the MPI calls are made by the master thread. This will happen if all MPI calls are outside OpenMP parallel regions or are in master regions. A thread can determine if it is the master thread with the MPI_Is_thread_main call. More precisely, it determines whether it is the same thread that called MPI_Init or MPI_Init_thread.
MPI_THREAD_SERIALIZED The MPI calls are made by only one thread at a time. This can be enforced in OpenMP by a construction like:

```
#pragma omp parallel
...
#pragma omp single
{
    ...MPI calls allowed here...
}
```

(as long as nested parallelism is not used.)
MPI_THREAD_MULTIPLE MPI calls can be made anywhere, by any thread.

All MPI implementations of course support MPI_THREAD_SINGLE. The nature of typical MPI implementations is such that they probably also support MPI_THREAD_FUNNELED, even if they don't admit it by returning this value from MPI_Init_thread, presuming they use a thread-safe malloc and other system calls, likely in any OpenMP application. Usually when people refer to an MPI implementation as "thread safe" they mean at the level of MPI_THREAD_MULTIPLE. It is worth noting that OpenMP encourages a style of programming that only requires MPI_THREAD_FUNNELED, so hybrid programming does not necessarily require a fully "thread safe" MPI.

3   Visualizing the Behavior of Hybrid Programs

Over the years we have found it surprisingly difficult to predict the behavior of complex MPI programs, and have developed a number of tools to assist in the process of understanding them as an important step in tuning them for performance. Understanding complex hybrid programs will be even more difficult. We describe here an extension to an existing tool to the hybrid case.

3.1   Jumpshot

Jumpshot [5,22] is a parallel program visualization program that we have long used to examine the detailed behavior of MPI programs. It provides a "Gantt chart" view of time lines of parallel processes, with colored rectangles to indicate the state of a process over a particular time interval. It also uses arrows from one line to another to indicate messages. While not scalable to very large numbers of processes (say, greater than 512), Jumpshot's panning and zooming capabilities, coupled with its summary views allowing a wide range of time scales to be viewed, have made it a valuable tool for studying the detailed behavior of parallel programs.

Jumpshot displays data from SLOG2 [14] files, which are written in an efficient way during the course of a parallel program execution. The library for logging events is provided by the MPE package [4] distributed with MPICH2 [6]. Both automatic logging of MPI calls via the MPI profiling interface and user-defined states and events are provided.

3.2   Jumpshot and Threads

Recently MPE, SLOG2 and Jumpshot have been extended to allow visualization of multi-threaded, and hence hybrid, programs. We made the assumption that the threads calling the MPI function or the MPE logging routines directly were POSIX `pthreads`, so that we could use the `pthread` library for the mutexes required as multiple threads wrote to the same memory buffer containing the logging records. MPE and SLOG2 were modified to include thread ID's in the log records.

Jumpshot needed to be augmented so that separate time lines for separate threads would be shown. Controls were added to the Jumpshot display to allow threads to be 1) collapsed into their parent processes, 2) grouped with their parent processes, or 3) grouped into separate communicators. (A common application structure in a program requiring `MPI_THREAD_MULTIPLE` is for separate threads in a process to use separate communicators.) The rest of the figures in this paper are screenshots of Jumpshot viewing SLOG2 files created by MPE logging of hybrid programs.

The first step in determining whether Jumpshot could be used with OpenMP was to determine whether the threads created by OpenMP compilers really were POSIX pthreads, which MPE, SLOG2, and Jumpshot had already been modified to handle. Fortunately, in all three of the environments described here, a simple diagnostic program, in which OpenMP threads created by `#pragma omp parallel` used the POSIX interface to request their `pthread` id's, was able to prove that this was the case, and that thread id's were reused in multiple parallel regions. This meant that no additional work was needed to adapt the MPE-SLOG2-Jumpshot pipeline to OpenMP. While most OpenMP implementations "park" threads like this, the situation could change for some emerging architectures that support fast thread creation.

Figure 1 is a Jumpshot view of a program with simultaneously communicating threads that demonstrates that MPICH2 is thread safe at the level of MPI_THREAD_MULTIPLE.

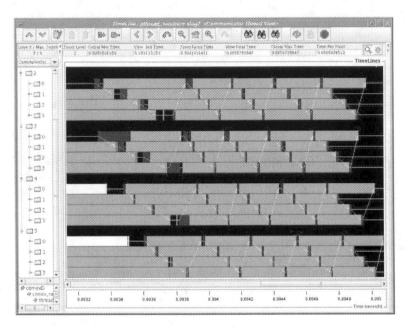

Fig. 1. Jumpshot's Communicator-Thread view of an MPI/pthreads program that forms a send and receive ring in the subcommunicator created from one single thread of each process

Other tools, such as Paraver [13,17], Vampir [11], and TAU [3] are also available for visualizing hybrid programs. We focus here on the use of Jumpshot because of its ability to show extreme detail and its wide availability, in particularly on the platforms presented here. Jumpshot is included as an optional viewer in recent releases of TAU.

4 The Hybrid Environments

The three machine/compiler environments that we tested were different in both hardware and software. All are quite new. The first is a "standard" Linux cluster based on AMD dual dual-core nodes. The latter two are examples of low-power, highly scalable computers. Our intention was not to explicitly compare the performance of these machines, but rather to demonstrate the viability of the hybrid programming model consisting of OpenMP and MPI on them, and to demonstrate the Jumpshot tool in these environments.

The details on the three platforms are as follows:

Linux cluster Each node is dual Opteron dual-core 2.8Ghz, i.e. four 2.8 Ghz
cores. The Intel 9.1 fortran compiler, `ifort`, is used with MPICH2-1.0.6p1
nemesis channel, which uses shared memory for intranode MPI communication and provides `MPI_THREAD_MULTIPLE` support. This is a typical Linux
cluster. Ours has multiple networks, but the experiments were done on Gigabit Ethernet.

IBM BG/P In IBM's BlueGene/P system [2], each compute node consists
of four PowerPC 850 Mhz cores. IBM's XLF 11.1 fortran cross-compiler,
`bgxlf_r`, is used with BlueGene MPI version V1R1M2 which provides
`MPI_THREAD_MULTIPLE` support. The BlueGene system has a high-
performance 3-D torus network for point-to-point communication.

SiCortex SC5832 The SiCortex SC5832[10] consists of 972 six-way SMP
compute nodes, i.e six MIPS 500 Mhz cores per node. The Pathscale 3.0.99
fortran cross-compiler, `scpathf95`, is used with the SiCortex MPI implementation, which provides `MPI_THREAD_FUNNELED` support. The SiCortex machine has a high-performance Kautz network.

All the Fortran compilers mentioned above provide OpenMP support for Fortran.
In addition, the companion C compilers provide OpenMP support for C.

5   Experiments

We did two sorts of experiments: first, a basic exploration of how things work,
and then two of the NAS parallel benchmarks, in order to investigate programs
amenable to the hybrid model for improving performance. The BG/P and SiCortex machines that we used are so new that they are still being tuned by the
vendors, so we focused on behavior rather than scalability of performance. Even
on the large machines we used a small number of nodes.

5.1   Basic Tests

We wrote a simple hybrid Fortran program and instrumented it with MPE. The
core of the program looks like this:

```
!$OMP PARALLEL DEFAULT(SHARED) PRIVATE(ii, jj, sum, ierr)
!$OMP DO
      do ii = 1, imax
         ierr = MPE_Log_event( blkA_startevt, 0, '' )
         sum = 0.0
         call random_number( frands )
         do jj = 1, jmax
            sum = sum + frands(jj) * jj
         enddo
         ierr = MPE_Log_event( blkA_finalevt, 0, '' )
      enddo
```

```
!$OMP END DO nowait
!$OMP END PARALLEL
```

This loop is repeated three times, to check consistency of thread ids, with three different event ids in the MPI_Log_event calls. The MPI calls are outside the parallel regions, so MPI_THREAD_FUNNELED is sufficient. Also, no MPI calls are made in the MPE_Log_event calls, so while pthread locks are used there, MPI_THREAD_MULTIPLE is not needed.

5.2   NAS Benchmarks

To initiate our study of hybrid programs, we choose a family of parallel benchmarks that have been written to take advantage of the hybrid style, the NAS parallel multi-zone benchmarks, NPB-MZ-MPI, version 3.1 [16], as the base code for our experiments. Two application benchmarks in NPB-MZ, namely BT and SP, were compiled with the respective OpenMP Fortran compiler on the AMD Linux cluster, the IBM Blue Gene P, and the SiCortex SC5832. We ran the codes in two different sizes (W and B) and different "modes" i.e., 16 MPI processes on four multi-core nodes and with four processes on four nodes, with each process having four threads (and with six each on the SiCortex). Note that the structure of NPB-MZ-MPI is such that it does not require MPI_THREAD_MULTIPLE.

From the outset it was clear that a tool capable of showing considerable detail would be useful in understanding what these programs were actually doing. For example, Figure 2 shows a short interval of execution on the Linux cluster. It is clear that although process 1 has 4 threads (actually 6, in this case) active at one time or another, something is preventing complete 4-way parallelism. (We would expect the red and green states to be stacked 4 deep, instead of offset in pairs the way they are.) This turned out to be a consequence of the way we had set the environment variables OMP_NUM_THREADS and NPB_MAX_THREADS, which had the side effect of deactivating the thread load-balancing algorithm in BT. Without Jumpshot we might not have realized that something was wrong.

Note that we can see that MPI communication is being done only by the first thread in each process. The "extra" threads that appear here are created by the load-balancing code in the BT benchmark, which overrides the OMP_NUM_THREADS environment variable set by the user. In later runs we controlled this with NPB_MAX_THREADS. Jumpshot alerted us to this anomaly. Results of our experiments are shown in Table 1.

The benchmarks were compiled with maximum optimization level known on each platform, i.e. -O3 with ifort on AMD Linux cluster, -O5 with bgxlf_r on BG/P and -O3 with scpathf95 on SiCortex. All these experiments are performed on 4 nodes on each chosen platform with either one process per core or one thread per core. For instance 16x1 refers to 16 processes running with 1 thread per process on 4 nodes, and 4x4 refers to 4 processes running with 4 threads per process on 4 nodes.

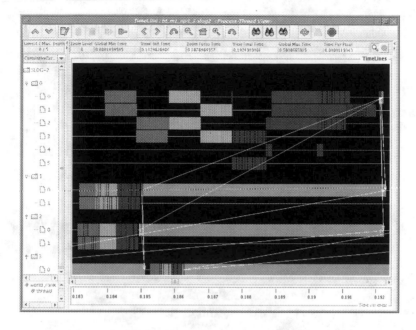

Fig. 2. Blocked Parallelism captured by Jumpshot

Table 1. NPB-MZ benchmark results are shown in seconds. The row labels are written in the form of <benchmark name>.<class name> .<process count> x <thread count>. Where benchmark name is either bt-mz or sp-mz, and class name is either W or B.

	AMD cluster	BG/P	SiCortex
bt-mz.W.16x1	1.84	9.46	20.60
bt-mz.W.4x4	0.82	3.74	11.26
sp-mz.W.16x1	0.42	1.79	3.72
sp-mz.W.4x4	0.78	3.00	7.98
bt-mz.B.16x1	24.87	113.31	257.67
bt-mz.B.4x4	27.96	124.60	399.23
sp-mz.B.16x1	21.19	70.69	165.82
sp-mz.B.4x4	24.03	81.47	246.76
bt-mz.B.24x1		·	241.85
bt-mz.B.4x6			337.86
sp-mz.B.24x1			127.28
sp-mz.B.4x6			211.78

Here are a few general observations.

– The hybrid approach provides higher performance on small (size W) version of BT on all three of these machines, as message-passing time dominated.
– For SP, even at size W, "MPI everywhere" was better.

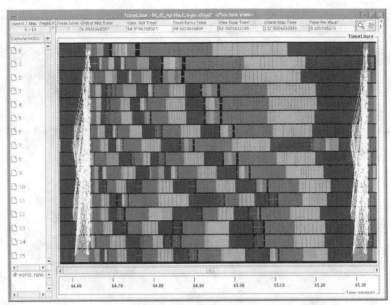

(a) np16x1, ViewDuration=0.7825s

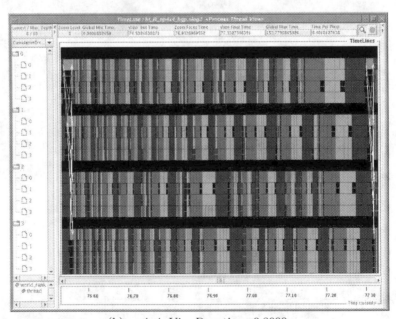

(b) np4x4, ViewDuration=0.8000s

Fig. 3. Jumpshot pictures of BT class B running on 4 BG/P nodes with either 1 process or 1 thread per core. (a) np16x1, ViewDuration=0.7825s. (b) np4x4, ViewDuration=0.8000s.

- On the size B problems, the "MPI everywhere" model was better than the hybrid approach.
- On the Sicortex only, we also ran 6 processes or threads per node, on 4 nodes, since each has 6 cores. The overall time dropped, showing the effect of applying more CPUs, but the machine still preferred the MPI everywhere model to the hybrid model.

Figure 3 shows the difference on BG/P between processes and threads. Subfigures a) and b) show similar time intervals.

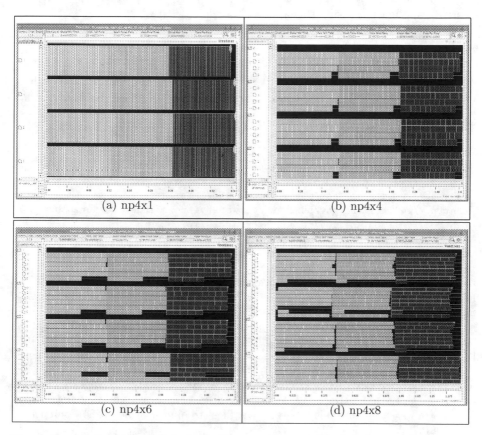

(a) np4x1   (b) np4x4

(c) np4x6   (d) np4x8

Fig. 4. Jumpshot pictures of the basic fortran program with OMP_NUM_THREADS=1, 4, 6, and 8 on 4 SiCortex nodes.

Figure 4 shows the effects of running various numbers of threads in processes on the SiCortex nodes, including more threads than cores, on our basic threading test program (not NPB). As the number of threads increases over the number of physical cores, the speed improvement due to parallelization of OpenMP threads does not seem to be diminishing yet. The reason could be because the SiCortex machine has yet to be fully optimized, so that extra CPU cycles are available for extra work, letting it appear to have more physical cores than there are.

6   Conclusions and Future Work

Our principal conclusion is that the hybrid programming model represented by the OpenMP and MPI standards is now available on the newest entries in the list of scalable high-performance computers as well as on traditional clusters. Jumpshot is no doubt only one of the tools available for inspecting the detailed behavior of hybrid programs, but so far few are both portable and freely available. Thus the pieces are in place, even on some of the largest and newest computers in the high performance computing complex, for application developers to create applications using this approach.

We have already begun extending the work presented here to other benchmarks at larger scale and to begin developing useful benchmarks specialized for the hybrid approach. Lack of space has limited us to the preliminary experiments presented here, but these show a promising beginning to a more thorough study.

References

1. http://crd.lbl.gov/~parry/hpcs_resources.html
2. http://www-03.ibm.com/servers/deepcomputing/bluegene.html
3. http://www.cs.uorigon.edu/research/tau
4. http://www.mcs.anl.gov/perfvis/download/index.htm#MPE
5. http://www.mcs.anl.gov/perfvis/software/viewers/index.htm#Jumpshot-4
6. http://www.mcs.anl.gov/research/projects/mpich2
7. http://www.mpi-forum.org/docs/docs.html
8. http://www.opengroup.org/onlinepubs/007908799/xsh/threads.html
9. http://www.openmp.org/blog/specifications
10. http://www.sicortex.com/products/sc5832
11. http://www.vampir.eu
12. Cappello, F., Etiemble, D.: MPI versus MPI+OpenMP on the IBM SP for the NAS benchmarks. In: Proceedings of SuperComputing 2000, IEEE Computer Society Press, Los Alamitos (2000)
13. Caubet, J., Gemenez, J., Labarta, J., De Rose, L., Vetter, J.: A Dynamic Tracing Mechanism for Performance Analysis of OpenMP Applications. In: Eigenmann, R., Voss, M.J. (eds.) WOMPAT 2001. LNCS, vol. 2104, Springer, Heidelberg (2001)
14. Chan, A., Lusk, E., Gropp, W.: An efficient format for nearly constant-time access to arbitrary time intervals in large trace files (to appear in Scientific Computing, 2008)
15. Chapman, B., Jost, G., van der Pas, R.: Using OpenMP. MIT Press, Cambridge (2007)
16. Van der Wijingaart, R.F., Jin, H.: the NAS parallel benchmarks, multi-zone versions. NAS Technical Report NAS-03-010, NASA Ames Research Center (2003), http://www.nas.nasa.gov/Resources/Software/npb.html
17. Freitag, F., Caubet, J., Labarta, J.: A trace-scaling agent for parallel application tracing. In: IEEE International Converence on Tools with Artivficial Intelligence (ICTAI), pp. 494–499 (2002)
18. Jost, G., Jin, H., an Mey, D., Hatay, F.F.: Comparing the openmp, mpi and hybrid programming paradigms on an SMP cluster. In: Proceedings of EWOMP (2003), http://www.nas.nasa.gov/News/Techreports/2003/PDF/nas-030019.pdf

19. Lusk, E., Yelick, K.: Languages for high-productivity computing: the DARPA HPCS language project. Parallel Processing Letters 17(1), 89–102 (2001)
20. Rabenseifner, R.: Hybrid parallel programming on HPC platforms. In: Proceedings of EWOMP (2003), http://www.compunity.org/events/ewomp03/omptalks/Tuesday/Session7/T01.pdf
21. Sterling, T., Messina, P., Smith, P.H.: Enabing Technologies for Petaflops Computing. MIT Press, Cambridge (1995)
22. Zaki, O., Lusk, E., Gropp, W., Swider, D.: Toward scalable performance visualization with Jumpshot. High Performance Computing Applications 13(2), 277–288 (1999)

First Experiences with Intel Cluster OpenMP

Christian Terboven, Dieter an Mey, Dirk Schmidl,
and Marcus Wagner

RWTH Aachen University, Center for Computing and Communication
Seffenter Weg 23, 52074 Aachen, Germany
{terboven,anmey,schmidl,wagner}@rz.rwth-aachen.de

Abstract. MPI and OpenMP are the de-facto standards for distributed-memory and shared-memory parallelization, respectively. By employing a hybrid approach, that is combing OpenMP and MPI parallelization in one program, a cluster of SMP systems can be exploited. Nevertheless, mixing programming paradigms and writing explicit message passing code might increase the parallel program development time significantly. Intel Cluster OpenMP is the first commercially available OpenMP implementation for a cluster, aiming to combine the ease of use of the OpenMP parallelization paradigm with the cost efficiency of a commodity cluster. In this paper we present our first experiences with Intel Cluster OpenMP.

1 Introduction

The main advantage of shared-memory parallelization with OpenMP over MPI is that data can be accessed by all instruction streams without reasoning whether it must be transferred beforehand. This allows for an incremental parallelization approach and leads to shorter parallel program development time. Complicated dynamic data structures and irregular and possibly changing data access patterns make programming in MPI more difficult, whereas the level of complexity introduced by shared-memory parallelization is lower in many cases. As OpenMP is a directive-based language, the original serial program can stay intact, which is an advantage over other shared-memory parallelization paradigms.

The downside of any shared-memory paradigm is that the resulting parallel program is restricted to execute in a single address space. Bus-based multiprocessor machines typically do not scale well beyond four processors for memory-intense applications. Larger SMP and ccNUMA systems require scalable and thus expensive interconnects. Because of that, several attempts to bring OpenMP to clusters have been made in the past.

In [6] an OpenMP implementation for the TreadMarks software has been presented, which supports only a subset of the OpenMP standard. In [7] an OpenMP implementation on top of the page-based distributed shared-memory (DSM) system SCASH has been presented for the Omni source-to-source translator. In this approach, all accesses to global variables are replaced by accesses into the DSM and all shared data is controlled by the DSM. Although the full OpenMP specification is implemented, support for the C++ programming language is missing.

R. Eigenmann and B.R. de Supinski (Eds.): IWOMP 2008, LNCS 5004, pp. 48–59, 2008.
© Springer-Verlag Berlin Heidelberg 2008

In 2006, Intel made the first commercial implementation of OpenMP for clusters available, named Intel Cluster OpenMP [4] (referred to as ClOMP in this paper). The full OpenMP 2.5 standard for Fortran, C and C++ is implemented, although nested parallel regions are not yet supported.

This paper is organized as follows: In section 2 we give an overview of OpenMP and Intel Cluster OpenMP. In section 3 we present micro-benchmark measurements of OpenMP and ClOMP constructs and discuss which types of applications we expect to profit from running on a cluster. In section 4 we present results of four applications utilizing Intel Cluster OpenMP. The current tool support for ClOMP is discussed briefly in section 5. We draw our conclusions and touch on future plans in section 6.

All measurements were carried out on a cluster of Fujitsu-Siemens Primergy RX200 servers equipped with two Intel Xeon 5450 (quad-core, 3.0 GHz) CPUs. All nodes are running Scientific Linux 5.1 and are connected via Gigabit Ethernet (referred to as Eth) and 4x DDR InfiniBand (referred to as IB). The InfiniBand adapters are attached to the PCI-Express bus. We used the Intel 10.1.011 compiler suite for 64-bit systems.

2 OpenMP

OpenMP consists of a collection of compiler directives, library functions and a few environment variables. It applies the so-called fork/join programming model.

As OpenMP is a shared-memory parallelization paradigm, all threads share a single address space, but still can have thread local storage to hold private data. It is the programmer's responsibility to control the scoping, that is the classification of variables into shared and private, of all variables that are used within a parallel region.

2.1 Memory Model

OpenMP provides a relaxed memory consistency model similar to the weak ordering memory model [3]. Each thread has a *temporary view* of the memory that is not required to be consistent with the memory at all times. Writes to memory are allowed to overlap other computation and reads from memory are allowed to be satisfied from a local copy of memory under some circumstances. For example, if within one synchronization period the same memory location is read again, this can be done from fast local storage (the temporary view, e.g. a cache). Thus, it is possible to hide the memory latency within an OpenMP program to some extent. This also allows Intel Cluster OpenMP to fulfill reads from local memory under certain circumstances, instead of accessing remote memory in all cases, as will be explained in the following subsection 2.2.

The *flush* construct of OpenMP serves as a memory synchronization operation, as it enforces consistency between the temporary view and the global view, by writing back a set of variables or even all thread's variables to the memory. All reads and writes from and to the memory are unordered with respect to each other (except for those being ordered by the semantics of the base language), but ordered with respect to an OpenMP flush operation. All explicit and implicit OpenMP barriers also contain an implicit flush operation.

2.2   Intel Cluster OpenMP

Beginning with version 9.1, the Intel C/C++ and Fortran compilers for Linux are available with Cluster OpenMP. The distributed shared-memory (DSM) system of Intel Cluster OpenMP is based on a licensed derivative of the TreadMarks software.

Intel has extended OpenMP with one additional directive: *sharable*. It identifies variables that are referenced by more than one thread and thus have to be managed by the DSM system. While certain variables are automatically made sharable by the compiler, some variables have to be declared sharable explicitly by the programmer, e.g. file-scope variables in C and C++. Thus, the programmer's responsibility for variable scoping has been extended to finding all variables that have to be made sharable, in the cases where the compiler is unable to detect it. As will be shown in section 4, this can sometimes be a tedious task for application codes.

For the Fortran programming language several compiler options exist to make different kinds of variables sharable automatically, e.g. all module or common block variables. In addition to finding all variables that have to be declared sharable, dynamic memory management in an application deserves some attention. For all variable allocations from the heap (e.g. by `malloc`), it has to be determined whether the memory should be taken from the regular heap, thus being only accessible by the thread calling `malloc`, or from the DSM heap, which is accessible by all threads. Intel Cluster OpenMP provides several routines to easily replace native heap memory management routines by DSM heap routines.

The task of keeping shared variables consistent across multiple nodes is handled by the Cluster OpenMP runtime library. Intel provides detailed information on how this process works in the product documentation and in the white paper [4]. In principle the mechanism relies on protecting memory pages via the `mprotect` system call; pages that are not fully up-to-date are protected against reading and writing. When a program reads from such a protected page, a segmentation fault occurs and after intercepting the corresponding signal the runtime library requests updates from all nodes, applies them to the page and then removes the protection. At the next access, the instruction finds the memory accessible and then the read will complete successfully. Still the page is protected against writing. In case of a write operation, a so-called twin page is created for further reads and writes on the accessing node, after the protection has been removed. The twin page then becomes the thread's temporary view.

The higher the ratio of cheap memory accesses, that means to thread private memory or to twin pages, versus expensive memory accesses, the better the program will perform. At each synchronization construct, e.g. a barrier, nodes receive information about pages modified by other nodes and invalidate those. As a consequence, the next access will be expensive.

3   Micro-benchmarks

In order to better understand the behavior of Intel Cluster OpenMP's DSM mechanism and to get an estimate of how expensive the DSM overhead is, we created a set of micro-benchmarks. In addition, we ported the well-known OpenMP micro-benchmarks [1] to Intel Cluster OpenMP and examined their performance on two different network fabrics.

Table 1 shows selected results of the OpenMP micro-benchmarks for traditional OpenMP and Intel Cluster OpenMP. The EPCC micro-benchmarks for OpenMP measure the overhead of OpenMP constructs by comparing the time taken for a section of code executed sequentially, to the time taken for the same code executed in parallel enclosed in a given directive. We ported the EPCC micro-benchmarks to Intel Cluster OpenMP by adding sharable directives, where necessary.

Table 1. Selected results (overhead in microseconds [us]) of the EPCC OpenMP micro-benchmarks for OpenMP and Intel Cluster OpenMP, with one thread per node

	OpenMP (1 node)	ClOMP (Eth)	ClOMP (IB)
PARALLEL FOR, 1 node	0.31	481.17	481.13
PARALLEL FOR, 2 nodes	1.00	1210.74	702.65
PARALLEL FOR, 4 nodes	1.12	1357.99	779.16
BARRIER, 1 node	0.01	480.59	480.71
BARRIER, 2 nodes	0.43	637.53	558.81
BARRIER, 4 nodes	0.60	662.70	634.64
REDUCTION, 1 node	0.35	481.54	482.44
REDUCTION, 2 nodes	1.54	1726.69	888.00
REDUCTION, 4 nodes	2.32	2202.01	1242.85

It is obvious that there is a severe difference in overhead between OpenMP and Intel Cluster OpenMP, independent of the network fabric. While for a run with a single thread only a small difference between the two network fabrics can be observed, the overhead increase with two and four threads is significantly lower on InfiniBand than on Ethernet. As will be seen in section 4, application codes resemble this behavior. We found that using a fast network like InfiniBand is crucial in order to exploit application scalability with Intel Cluster OpenMP.

We implemented a couple of own micro-benchmarks to test the DSM performance by employing the same measurement approach as the EPCC micro-benchmarks:

- *testheap*: A number of pages is allocated via `kmp_aligned_sharable_malloc` (OpenMP: `valloc`), then they are written and then freed again. This process is repeated a couple of times and the average runtime is calculated.
- *read_f_other*: The time required to read a page allocated via the DSM by a different thread is measured. For the Cluster OpenMP runtime that requires transferring the page.

- *write_t_other*: Similar to read_f_other, but now the page allocated by a different thread is written. For the Cluster OpenMP runtime that requires creating a twin page.

The performance results for traditional OpenMP and Intel Cluster OpenMP are shown in table 2. The OpenMP measurements were run with two threads. Both ClOMP measurements were run with two Cluster OpenMP threads schedules on one or two nodes.

Table 2. Selected results (two threads, overhead in microseconds [us]) of our Cluster OpenMP micro-benchmarks

	testheap	read_f_other	write_t_other
OpenMP	2.76	1.6	2.11
ClOMP (Eth), 1 node	6.79	1.75	2.48
ClOMP (Eth), 2 nodes	4.77	247.24	251.13
ClOMP (IB), 1 node	4.37	1.78	2.49
ClOMP (IB), 2 nodes	22.96	94.94	94.38

Obviously, allocating dynamic memory is more expensive with ClOMP. With Intel Cluster OpenMP, special care has to be taken in case of dynamic data structures which involve many allocations, maybe even hidden from the user via an abstract interface.

Although Cluster OpenMP allows the programmer to access memory on other nodes transparently, from a performance perspective this is not for free. Intel Cluster OpenMP can be started to use more than one thread per node, instead of multiple processes on one node. In that case, accessing memory from a different thread on the same node is significantly cheaper.

With Intel Cluster OpenMP it is even more important to respect and stick to the following OpenMP tuning advices:

- *Enlarge the parallel region*: Creating a team of threads at the entrance to a parallel region and putting it aside at the exit involves some overhead, although most current compilers do a good job in keeping it minimal. Fewer and shorter serial parts contribute to better scalability, thus parallel regions should be as large as possible in most cases. With Cluster OpenMP the overhead of creating or activating a team of threads is higher than for OpenMP, as all involved nodes have to communicate.
- *Work on data locally*: Keeping data local is very important on ccNUMA architectures. We found that tuning measures for ccNUMA also improve performance on Cluster OpenMP. If threads are accessing local memory, no page transportation occurs and no twin page has to be created.
- *Prevent false sharing*: Normally, false sharing occurs when threads write to different parts of the same cache line, e.g. if two threads are running on two different cores that do not share a cache, then only one core can hold the valid cache line, the other core has to wait to perform any update. This

can affect the performance significantly. In the case of Cluster OpenMP, false sharing becomes an issue on a per page basis. If two or more threads write to different locations on the same page, the update process has to occur at the next synchronization point. This kind of problem is pretty hard to detect in complex applications, but in many cases can be resolved by inserting appropriate padding.

4   Applications

4.1   Jacobi

We tried Intel Cluster OpenMP on the Jacobian solver available on the OpenMP website. We measured the scalability using a matrix size of 6000 × 6000 and 100 iterations. We compared the ClOMP version to traditional OpenMP and two MPI implementations, one with synchronous communication and one with asynchronous communication. Selected results are shown in figure 1.

In all versions the domain decomposition approach for the parallelization is exactly the same. With MPI the data on the boundary has to be transferred explicitly, while the OpenMP programmer does not need to care. As the DSM system of Intel Cluster OpenMP works on a per page basis, in some cases depending on the total number of threads and the number of threads per node, some threads will have to access pages on other nodes for reading data at or near the boundaries. This can be prevented by appropriate padding.

We found that binding Cluster OpenMP threads to scattered cores improves the performance of up to 10% and was most effective for the runs with two threads per node.

It can be noticed that the scalability on one 8-core node is limited to four threads, as the Jacobi solver stresses the memory bandwidth. Thus, running with two Cluster OpenMP threads on two nodes shows a better scalability (1.92 over 1.67) than the traditional OpenMP version on a single node, as the memory bandwidth available to the application is virtually doubled by running on two nodes. Of course, using more than four threads per node does not improve scalability with Cluster OpenMP for the same reason. Instead, the performance drops slightly. The maximum measured speedup for ClOMP on 16 nodes is 12.68 with InfiniBand and four threads per node and 9.51 with Ethernet and two threads per node.

Overlapping communication and computation with asynchronous MPI is particularly beneficial when employing the slower Gigabit Ethernet network fabric, as shown for the MPI case with one process per node, whereas for InfiniBand it does not make a big difference. Likewise the Cluster OpenMP version profits from the faster network, because communication and computation cannot be overlapped explicitly. In all cases MPI clearly outperforms Cluster OpenMP. Both MPI versions deliver a speedup of about 27 with sixteen nodes and four processes per node when using the fast InfiniBand network fabric.

There are two places in the program where communication is involved: In updating data on the boundaries of the subdomains and in the reduction operation to

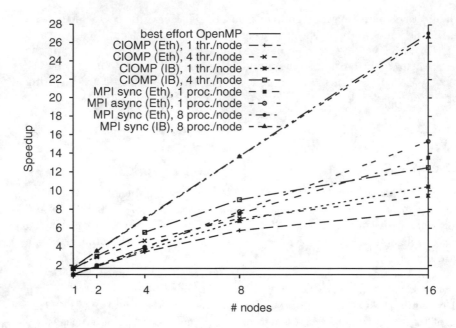

Fig. 1. Speedup of the Cluster OpenMP and MPI versions of Jacobi

calculate the error estimation. In order to improve the Cluster OpenMP version, we implemented prefetching with an additional Posix-thread, that would be similar to overlapping computation and communication. Unfortunately, we were unable to achieve any significant performance improvement by prefetching the boundary data. To understand this result, we estimated the runtime for eight ClOMP threads on eight nodes assuming perfect scalability and the runtime for page transfers and reduction operations on the basis of our micro-benchmark measurements. As both estimations only differ within 1.5 percent, we concluded that with prefetching there is only little to gain. This result corresponds to the observation that the asynchronous and synchronous MPI versions perform similarly on the fast InfiniBand network. Applying the prefetch strategy in combination with the slower GE network, we observed a slight speedup improvement of about four percent.

As our micro-benchmark experiments revealed that an MPI reduction operation performs significantly better than a reduction operation in Cluster OpenMP, we linked the Intel Cluster OpenMP program with the Intel MPI library and called the MPI reduction operation from within the Cluster OpenMP program - an approach which is not officially supported by Intel. By replacing Cluster OpenMP's reduction operation with the MPI reductions, we got an increase in speedup of only 1.5% on the presented dataset, as we had to introduce additional locking. The improvement of course depends on the number of reduction operations called. We concluded that there is still room for improvement in the Intel Cluster OpenMP implementation, as the reduction can be implemented more efficiently than in our experiments.

4.2   Sparse Matrix-Vector-Multiplication

A sparse matrix-vector-multiplication (SMXV) typically is the most time consuming part in iterative solvers. In order to estimate whether Intel Cluster
OpenMP is suited for this class of applications, we examined the SMXV benchmark kernel of DROPS, a 3D CFD package for simulating two-phase flows with
a matrix of some 300 MB and about 19,600,000 nonzero values.

Table 3. Performance of SMXV [MFLOP/s] and runtime of GMRES [seconds]

	SMXV-rows		SMXV-nonzeros		GMRES	
	1 thread p. node	6 threads p. node	1 thread p. node	6 threads p. node	1 thread p. node	4 threads p. node
OpenMP UMA	502.9	976.6	501.6	998.1	167.9	94.7 67.9 (8 threads)
OpenMP ccNUMA	326.3	793.9	324.5	1147.6	n.a.	n.a.
ClOMP IB, 1 node	494.6	840.3	501.6	926.0	167.9	95.4
ClOMP IB, 4 nodes	290.8	443.7	1900.6	3047.4	71.1	57.8
ClOMP IB, 8 nodes	281.8	277.7	3580.3	5113.9	87.6	70.1
ClOMP IB, 16 nodes	285.7	196.4	6111.2	9922.2	115.1	104.4

The performance of the SMXV benchmark is shown in table 3. In addition to
the Harpertown-based systems (UMA), we evaluated the performance on a Sun
Fire V40z server system, equipped with four AMD Opteron 848 single-core 2.2
GHz CPUs (ccNUMA), which provides a ccNUMA architecture. We compared
two parallelization strategies: In the *rows*-strategy the parallel loop runs over
the number of rows and a dynamic loop schedule is used for load balancing,
while in the *nonzeros*-strategy the number of nonzeros is statically partitioned
into blocks of approximately equal size, one block for each thread.

The *nonzeros*-strategy outperforms the *rows*-strategy on the ccNUMA architecture and on Intel Cluster OpenMP as well, when carefully initializing all data
while respecting the operating system's first touch policy. Though the dynamic
loop scheduling in the *rows*-strategy provides good load balance, the memory
locality is not optimal. The *nonzeros*-strategy shows a negligible load imbalance
for the given dataset, but its advantage is that each thread works on local data.
Using the segvprof.pl tool we observed the number of page transfers dropping
by a factor of 50. For this strategy, there is only little difference between Gigabit Ethernet and InfiniBand, as there is only little communication involved. In
short, Cluster OpenMP behaves like a distinct ccNUMA architecture.

4.3   GMRES Solver

We also examined the GMRES-type iterative solver of the DROPS application and used the same matrix setup as for the SMXV kernel. The performance of the GMRES benchmark is also shown in table 3.

The best speedup with traditional OpenMP is 1.8 with four threads per node and 2.5 with eight threads per node, reducing the runtime from 167.9 to 67.9 seconds. The scalability of this code is limited by the available memory bandwidth as well as by the synchronization points in each iteration.

The original OpenMP version did not scale at all after being ported to Intel Cluster OpenMP. As the vectors spanning up the Krylov-subspace were shared, clearly too many page transfers occurred, thus we privatized those vectors for the ClOMP version. Unluckily, we were then unable to employ 8 ClOMP threads per node because of limitations of the sharable heap, as replicating the vectors for each thread increased the memory consumption significantly. The best effort speedup with Intel Cluster OpenMP is 2.9 with four threads per node on four nodes, thus 16 threads in total. Still the scaling of the ClOMP version does not fulfill our expectations. Computing the next Krylov-subspace vector involved page transfers and that amount increases with the number of threads and nodes used. We found that this part of the GMRES kernel is solely responsible for the performance dropdown with more than four nodes.

Our future plan is to re-design the GMRES kernel to be better suited for Cluster OpenMP by sharing the Krylov-subspace vectors, but optimizing the access pattern throughout the algorithm.

4.4   PANTA

PANTA is a 3D solver that is used in the modeling of turbomachinery [9]. The package used in our experiments consists of about 50,000 lines of Fortran 90 code. Several approaches to parallelize this code have been described, e.g. [5]. In order to achieve the best possible speedup with Cluster OpenMP, we have chosen the highest parallelization level currently exploited with OpenMP, that is a loop over 80 inversion zones.

We had to manually compute the distribution of loop iterations onto threads, as the OpenMP DO work-sharing construct was not applicable in this case because of the code structure. As the number of loop iterations is relatively small and at the end of each loop iteration there is a critical region in which some global arrays are updated in a reduction-type manner, we cannot expect good scaling from this code.

Creating a Cluster OpenMP version of the PANTA code parallelized with OpenMP was straight forward: We enabled the compiler's autodetection and propagation of sharable variables and asked the compiler to make all argument expressions, all common block variables, all module variables and all save variables sharable by default.

We are aware of the fact that making all these variable types sharable by default puts more variables under the control of the DSM than necessary and that this will probably cause a performance penalty. Nevertheless, the scalability of the Cluster OpenMP version on a single node is similar to the OpenMP version. Better scalability with traditional OpenMP on a single node is prohibited because the available memory bandwidth is saturated. Using Intel Cluster OpenMP, we can use more than one node and thus effectively increase the available memory bandwidth. Using two nodes, the best effort speedup can be increased from 2.9 with traditional OpenMP to 3.3, using four nodes to 4.3. Adding more nodes will only lead to slight improvements. For PANTA, Gigabit Ethernet performs worse than traditional OpenMP in all cases.

Unluckily, the current version of Intel Cluster OpenMP does not support Nested OpenMP. For the PANTA code, there is an additional OpenMP parallelization at the loop level available, namely at the linear equation solver [5]. We suspect that employing this level with two or even four threads per node would increase the total scalability of the program.

4.5 Fire

The Flexible Image Retrieval Engine (FIRE) [2] has been developed at the Human Language Technology and Pattern Recognition Group of the RWTH Aachen University. Given a query image and the goal to find k images from a database that are similar to the query image, a score is calculated for each image from the database. In [8] two layers have been parallelized with OpenMP and displayed nearly linear scalability. Shared-memory parallelization is obviously more suitable than distributed-memory parallelization for the image retrieval task, as the image database can then be accessed by all threads and does not need to be distributed. Because of that, we expected FIRE to be a perfect candidate for Intel Cluster OpenMP as searching through the database involves very little synchronization and only negligible writing to shared memory.

To make variables of the C++ STL sharable, instances of such variables have to use the `kmp_sharable` allocator. In order to achieve this, that allocator has to be specified at the variable declaration. On one hand this solution is elegant and does not require many code changes at the declaration point, but on the other hand the type signature of the variable is changed. This implies that if such a variable is passed as a parameter to a function, the function declaration has to be changed to reflect the type change.

The FIRE code makes extensive use of the STL. In order to make FIRE work with Intel Cluster OpenMP, virtually the whole code base would have to be touched and nearly every class would have to be changed. This is not feasible in a limited amount of time and in contrast to the findings in [8] that with OpenMP only very little code changes were necessary. Providing an STL which allocates all STL variables on the DSM heap might be a solution for this and similar codes.

5 Tool Support

The DSM-mechanism used by Intel Cluster OpenMP uses segmentation fault signals to activate the page movement and synchronization mechanism. That makes debugging a Cluster OpenMP program very hard, if not impossible, if the debugger cannot be taught to ignore the segfaults and to not step into the Cluster OpenMP library's handler routine. In doing so we successfully used the Intel command line debugger and the TotalView GUI-based debugger with Cluster OpenMP programs. Nevertheless, using traditional debuggers is not very helpful in finding errors related to Intel Cluster OpenMP. The typical problem is that a variable has erroneously not been made sharable. In this case some threads will run into segmentation faults when accessing that memory location, but the runtime system is unable to deliver the page and thus terminates the program in most cases.

In order to find the places in which accesses to variables that are not sharable occur, one can use the command line tool `addr2line` on a core dump. We found it easy to use and in most cases it was no problem to figure out which variable caused the problem. Intel has announced that future versions of the Intel Thread Checker tool will also find variables that should be made sharable.

In addition, Intel delivers a command line tool named `segvprof.pl` that provides means to count the number of segmentation faults on the function level. This can be handy in locating parts of the program that are not performing well, as e.g. too many accesses to remote pages occur.

Again, this tool is very basic in it's current form and for complex codes like PANTA, the provided functionality is too limited to find and understand performance problems related to Cluster OpenMP. Intel has announced that future versions of the Intel Trace Collector and Analyzer will support such an analysis.

6 Conclusions and Future Work

Intel Cluster OpenMP allows shared-memory OpenMP programs to be executed on a cluster. It takes advantage of the relaxed consistency memory model of OpenMP. Nevertheless, OpenMP primitives get in average two to four orders of magnitudes more expensive.

Intel Cluster OpenMP proved to be successful for several small applications and while preserving the easier and more comfortable parallelization paradigm of OpenMP and shared-memory, a cluster of SMP nodes could be exploited. But for more complex applications like the GMRES-kernel and PANTA, scalability does not come for free and further tuning effort has to be invested.

We ran into problems with C++ programs employing the STL, which still have to be resolved. We suspect that there is room for improvement concerning Intel's current implementation of reductions and on the tool support.

Future work will be to evaluate more programs of the scientific domain with Intel Cluster OpenMP. Algorithms like GMRES have to be re-designed to some extent to respect the high latencies for remote accesses with ClOMP. We will apply tuning measures to Cluster OpenMP programs: Porting codes like PANTA was straight forward because of the compiler features provided, still the full performance potential has not yet been achieved. We are interested in combining Cluster OpenMP with other parallelization paradigms to enable multi-level parallelism.

Acknowledgements

We sincerely thank Jay Hoeflinger and Larry Meadows from Intel for providing hints on potential performance improvements.

References

1. Bull, J.M.: Measuring Synchronisation and Scheduling Overheads in OpenMP. In: European Workshop on OpenMP (EWOMP), Lund, Sweden (September 1999)
2. Deselaers, T., Keysers, D., Ney, H.: Features for Image Retrieval: A Quantitative Comparison. In: Rasmussen, C.E., Bülthoff, H.H., Schölkopf, B., Giese, M.A. (eds.) DAGM 2004. LNCS, vol. 3175, pp. 228–236. Springer, Heidelberg (2004)
3. Hennessy, J.L., Patterson, D.A.: Computer Architecture - A Quantitative Approach. Morgan Kaufmann Publishers Inc, San Francisco (2006)
4. Hoeflinger, J.P.: Extending OpenMP to Clusters (2006)
5. Lin, Y., Terboven, C., an Mey, D., Copty, N.: Automatic Scoping of Variables in Parallel Regions of an OpenMP Program. In: Chapman, B.M. (ed.) WOMPAT 2004. LNCS, vol. 3349, pp. 83–97. Springer, Heidelberg (2005)
6. Lu, H., Hu, Y.C., Zwaenepoel, W.: OpenMP on Network of Workstations (1998)
7. Sato, M., Harada, H., Hasegawa, A., Ishikawa, Y.: Cluster-enabled OpenMP: An OpenMP compiler for the SCASH software distributed shared memory system. Scientific Programming 9(2,3), 123–130 (2001)
8. Terboven, C., Deselaers, T., Bischof, C., Ney, H.: Shared-Memory Parallelization for Content-based Image Retrieval. In: ECCV 2006 Workshop on Computation Intensive Methods for Computer Vision (CIMCV), Graz, Austria (May 2006)
9. Volmar, T., Brouillet, B., Gallus, H.E., Benetschik, H.: Time Accurate 3D Navier-Stokes Analysis of a 1.5 Stage Axial Flow Turbine (1998)

Micro-benchmarks for Cluster OpenMP Implementations: Memory Consistency Costs

H.J. Wong, J. Cai, A.P. Rendell, and P. Strazdins

Australian National University,
Canberra ACT 0200, Australia
{jin.wong,jie.cai,alistair.rendell,peter.strazdins}@anu.edu.au

Abstract. The OpenMP memory model allows for a temporary view of shared memory that only needs to be made consistent when `barrier` or `flush` directives, including those that are implicit, are encountered. While this relaxed memory consistency model is key to developing cluster OpenMP implementations, it means that the memory performance of any given implementation is greatly affected by which memory is used, when it is used, and by which threads. In this work we propose a micro-benchmark that can be used to measure memory consistency costs and present results for its application to two contrasting cluster OpenMP implementations, as well as comparing these results with data obtained from a hardware supported OpenMP environment.

1 Introduction

Micro-benchmarks are synthetic programs designed to stress and measure the overheads associated with specific aspects of a hardware and/or software system. The information provided by micro-benchmarks is frequently used to, for example, improve the design of the system, compare the performance of different systems, or provide input to more complex models that attempt to rationalize the runtime behaviour of a complex application that uses the system.

In the context of the OpenMP (OMP) programming paradigm, significant effort has been devoted to developing various micro-benchmark suites. For instance, shortly after its introduction Bull [3,4] proposed a suite of benchmarks designed to measure the overheads associated with the various synchronization, scheduling and data environment preparation OMP directives. Other notable OMP related work includes that of Sato et al. [16] and Müller [12].

All existing OMP micro-benchmarks have been developed within the context of an underlying hardware shared memory system. This is understandable given that the vast majority of OMP applications are currently run on hardware shared memory systems, but it is now timely to reassess the applicability of these micro-benchmarks to other OMP implementations. Specifically, for many years there has been interest in running OMP applications on distributed memory hardware such as clusters [14,8,9,2,7]. Most of these implementations have been experimental and of a research nature, but recently Intel released a commercial product that supports OMP over a cluster – Cluster OpenMP (CLOMP) [6]. This

R. Eigenmann and B.R. de Supinski (Eds.): IWOMP 2008, LNCS 5004, pp. 60–70, 2008.

interest, plus the advent of new network technologies that offer exceptional performance and advanced features like Remote Direct Memory Access (RDMA), stands to make software OMP implementations considerably more common in the future. Also it is likely that the division between hardware and software will become increasingly blurred. For example, in recent work Zeffer and Hagersten [19] have proposed a set of simplified hardware primitives for multi-core chips that can be used to support software implemented inter-node coherence.

This paper addresses the issue of developing a benchmark to measure the cost of memory consistency in cluster OMP implementations. In the OMP standard, individual threads are allowed to maintain a temporary view of memory that may not be globally consistent [13]. Rather, global consistency is enforced either at synchronization points (OMP `barrier` operations) or via the use of the OMP `flush` directive[1]. On a hardware shared memory system both operations are likely to involve hardware atomic and memory control instructions such as the `fence`, `sync`, or `membar` operations supported on x86, POWER, and SPARC architectures respectively. These operations can be used to ensure that the underlying hardware does not a) move any of the load/store operations that occurred before the barrier/flush to after or vice versa, and b) ensure that all load/store operations that originated before the barrier/flush are fully completed (e.g. not waiting on a store queue).

It is important to note that memory consistency and cache coherency are different concepts. Cache coherency policies ensure that there is a single well defined value associated with any particular memory address. Memory consistency on the other hand determines when a store instruction executed by one thread is made visible to a load operation in another thread and what that implies about other load and store operations in both threads [1].

A fundamental difference between hardware and software supported OMP is that much of the work associated with enforcing the OMP memory consistency model occurs naturally in the background on a hardware shared memory system, but this is not the case for most software supported OMP systems. That is on a typical hardware shared memory system the result of a store operation becomes globally visible as soon as it is propagated to a level where the cache coherency protocol knows about it (e.g. when it has been retired from the store queue). Thus if an OMP program has one thread executing on one processor that is periodically changing the value of a shared variable, while another thread is executing on another processor and is periodically reading the same variable, the reading thread is likely to see at least some of the different values written by the other thread, and this will be true regardless of whether there are any intervening OMP `barrier` or `flush` directives. This is not, however, required by the OMP memory consistency model (and must not be relied upon by any valid OMP program). From the OMP perspective it can be viewed as if the cache coherency hardware is speculatively updating global shared memory on

---

[1] Note that OpenMP `flush` operations may be implied. For instance, they are implied before reading from and after writing to `volatile` variables. Thus, the temporary and global views of these variables are kept consistent automatically.

the expectation that the current change will not be overwritten before the next `barrier` or `flush` directive is encountered; and if the change is overwritten, propagating it will have been a waste of time and bandwidth!

For software supported OMP systems, propagating changes to global memory generally implies significant communication cost. As a consequence global updates are generally stalled as long as possible, which for OMP means waiting until a thread encounters a `flush` or `barrier` directive. What happens then varies greatly between different cluster OMP implementations. The aim of this paper is to outline a micro-benchmark that can be used to quantify these differences, and then to illustrate its use on two software enabled OMP systems, comparing the results with those obtained from a hardware supported OMP system.

The following section outlines the memory consistency benchmark, while Section 3 details the two software supported OMP systems used in this work. Section 4 contains the results obtained for the two software OMP systems and the contrasting hardware supported OMP system. Finally Section 5 contains our conclusions and comments for future work.

2   MCBENCH: A Memory Consistency Benchmark

The goal of the Memory Consistency Benchmark (MCBENCH)[2] is to measure the overhead that can be attributed to maintaining memory consistency for an OMP program. To do this, memory consistency work is created by *first* having one OMP thread make a change to shared data and then flush that change to the globally visible shared memory; and *then* having one or more other OMP threads flush their temporary views so that the changes made to the shared data are visible to them.

As noted above it is important that the readers' flushes occur *after* the writer's flush, otherwise OMP does not require the change to have been propagated. Both these requirements are met by the OMP `barrier` directive since this contains both synchronization and implicit flushes [13]. Accordingly, the general structure used by MCBENCH is a series of change and read phases that are punctuated by OMP `barrier` directives (where implicit flushes and synchronization occurs) to give rise to memory consistency work.

Since the above includes other costs that are not related to the memory consistency overhead, it is necessary to determine a reference time. This is done by performing the exact same set of operations but using private instead of shared data. The difference between the two elapsed times is then the time associated with the memory consistency overhead.

To ensure that the same memory operations are performed on both the private and shared data, the MCBENCH kernel is implemented as a routine that accepts the address of an arbitrary array. Figure 1 shows that this array of a bytes is divided into chunks of fixed size c which are then assigned to threads in a round-robin fashion. In the Change phase, each thread changes the bytes in

---

[2] MCBENCH is available for download at `http://ccnuma.anu.edu.au/dsm/mcbench`.

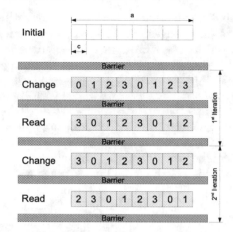

Fig. 1. MCBENCH – An array of size a-bytes is divided into chunks of c-bytes. The benchmark consists of Change and Read phases that can be repeated for multiple iterations. Entering the Change phase of the first iteration, the chunks are distributed to the available threads (four in this case) in a round-robin fashion. In the Read phase after the barrier, each thread reads from the chunk that its neighbour had written to. This is followed by a barrier which ends the first iteration. For the subsequent iteration, the chunks to Change are the same as in the previous Read phase. That is, the shifting of the chunk distribution only takes place when moving from the Change to Read phases.

their respective chunks. This is followed by a barrier, and the Read phase where the round-robin distribution used in the Change phase is shifted such that, had the array been a shared one, each thread will now read the chunks previously changed by their neighbours. The size of the shared array is the total number of bytes that was modified during the Change phase, and this is also the total number of modified bytes that must be consistently observed in the subsequent Read phase. Thus, this number represents the memory consistency workload in bytes that the underlying memory system must handle.

3   Two Software Supported OMP Implementations

Two software supported OMP implementations have been used in this work; both layer OMP over a page-based software Distributed Shared Memory (DSM) system [15,14,8,9,7]. The first is Intel's Cluster OpenMP product (CLOMP) [6] that uses a modified form of the TreadMarks DSM [10] for its underlying shared memory. The second is based on the Omni OpenMP Compiler [11] and uses the SCLIB DSM that has been developed by one of us [18]. The latter stands for "SCash LIBrary" and is so named because the consistency model used is the same as SCASH [5], but in contrast to SCASH it uses MPI for its communication requirements as opposed to the specialized PM communication library [17].

Two key activities that DSMs need to perform are detecting access to shared data and determining what changes, if any, have been made to local copies of shared data so that those changes can be made globally visible:

- **Detecting Shared Data Accesses** – At the granularity of a page, a subset of the virtual address space associated with each user process is designated as globally shared. Access to these shared pages is detected by using `mprotect` to modify the page's memory protection and implementing a handler for the associated `SIGSEGV` signal. To catch both read and write accesses, the `PROT_NONE` setting can be used, while to catch only writes `PROT_READ` is used.
- **Determining Changes to Shared Data** – When a write is attempted on a read-only page, a copy of the page is made before modifying the memory protection to allow the write to proceed. This process is called *twinning*. The presence of the twin then makes it possible to determine if changes have been made to any data on that page at a later stage. This process is called *diffing*. Using the *twinning-and-diffing* strategy makes it possible to have multiple threads modifying the same page concurrently.[3]

TreadMarks and SCLIB differ in that SCLIB is a home-based DSM while TreadMarks is not (or homeless). Home-based means that for each page of globally shared memory there is a uniquely defined node that is responsible for maintaining the currently valid version of that page, i.e. the *master copy*. An advantage of the home-based approach is that a thread only needs to communicate with one node in order to obtain a fresh copy of any given shared page or to send *diffs* relating to that page. In contrast, TreadMarks does not maintain any master copy of pages. Rather, *diffs* are held by their respective writers and only requested for when required. The advantage of this approach is that if a page is only accessed by one thread, there would be no communication of *diffs* pertaining to that page. This makes the DSM naturally adaptive to data access phase changes within the user application.

The difference between the two approaches is succinctly captured by what happens in a barrier. In TreadMarks, *diffs* are created for all modified pages which are then stored in a repository. Write notices are communicated amongst the threads, detailing which pages have been modified and by which threads. Pages that have been modified are also invalidated by changing the page protection to `PROT_NONE`. Post barrier, any access made to a modified page will invoke the `SIGSEGV` handler that fetches diffs from the various writer threads and applies them to the local page, thereby updating the page. At this point the page can be made read-only.

SCLIB is similar to TreadMarks in that *diffs* for all the modified pages are made during the barrier. However, rather than storing these in repositories, the *diffs* are communicated immediately to the page-home and applied directly onto the *master copy* of the page. Thus, at the end of the barrier, all master copies of

---

[3] Although multiple writers can modify the same page concurrently, these should be on separate portions of that page, otherwise the result from merging the diffs will be non-deterministic.

Table 1. Details of the experimental environments used in this work

Item	Detail	GigE Cluster	InfiniBand Cluster	Hardware (Sun V1280)
CPU	Manufacturer	AMD	Intel	Sun
	Type	Athlon 64 X2 4200	Xeon 5150	UltraSPARC
	Clock	2.2GHz	2.66GHz	900MHz
	Cores	2	2	1
Nodes	Count	8	20	3 (boards)
	CPU Sockets	1	2	4 (per board)
Network	Type	Gigabit Ethernet	InfiniBand (4x DDR)	-
	Latency	60 usec	3 usec	-
	Bandwidth	98 MB/s	1221 MB/s	-
OpenMP	Impl1	ICC-10.0.026, CLOMP-1.3, -O	ICC-10.0.026, CLOMP-1.3, -O	-
	Impl2	Omni/SCLIB, OpenMPI-1.1.4 (GCC-3.3.5), -O	Omni/SCLIB, OpenMPI-1.2.5 (GCC-4.1.2), -O	-
	Impl3	-	-	Sun Studio 11, -xopenmp=noopt

pages will be up-to-date. Post barrier, a page is read-only for the home thread and either read-only or invalid for all other threads, depending on whether there were changes[4]. Subsequent page faults cause the contents of a page to be fetched from the page-home, updating the temporary view of the faulting thread.

In summary, TreadMarks only sends the locally made changes to other threads when and if those other threads request them [5]; this restricts communication within the barrier to relatively short messages that just notify the other nodes that something has changed in a given page. SCLIB, on the other hand, sends the actual changes to a page to the home thread where the master copy is updated during the barrier event. In principle the time spent within the TreadMarks barrier is shorter than for SCLIB, but this comes at the expense of possibly having to retrieve *diffs* from multiple threads after the barrier and apply them. *A priori* it is not easy to judge which scheme is best.

4   Experimental Results

MCBENCH was used to measure the memory consistency overheads of the two software OMP implementations introduced in the previous section on the two different cluster environments detailed in Table 1. Some results were also obtained using the Sun Studio 11 OMP compiler on a Sun V1280 hardware shared memory machine.

---

[4] In a single-writer situation, the writer, if not the home of the page, will not receive an invalidation notice because its copy is as up-to-date as the master copy.

[5] Or when the repository of diffs become too full.

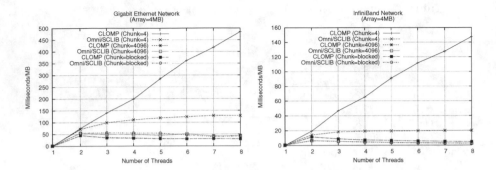

Fig. 2. MCBENCH results using an array of 4MB on the Gigabit Ethernet and Infini-Band clusters. The chunk size is 4 bytes, 4096 bytes, and blocked (4MB/n).

There are three parameters that can be adjusted when running MCBENCH. These are the size of the array (a), the chunk size (c), and the number of threads (n). The first parameter, a, represents the OMP memory consistency workload at each iteration in terms of bytes. This array is then divided into chunks of c bytes, which the available n threads then modify and read using the round-robin distribution illustrated in Figure 1. Thus, varying the chunk size modifies the memory access patterns of the threads.

4.1   Analysis: Software OMP Implementations

In studying the software OMP implementations, we observe how the overheads of the implementations scale with the number of threads. Scalability of this nature is important for clusters because the primary means of increasing computational power in a cluster is to add more nodes.

The chunk sizes that were used are 4 bytes, 4096 bytes, and *blocked* ($\lfloor \frac{a}{n} \rfloor$ bytes). At 4 bytes, the memory access pattern represents the worst case for the two software OMP implementations because they are layered on page-based DSMs whose pages are many times larger than $4n$ bytes. The result of such an access pattern is that each thread has to perform memory consistency over the whole array even though it only modifies/reads $\frac{a}{n}$ bytes in a phase. The second chunk size is 4096 bytes and is the same as the page size of the DSMs used by both software OMP implementations. By further aligning the shared array to the page line, a dramatic difference can be observed in the overheads. Lastly, the *blocked* chunk size is a simple block distribution of the shared array between the available threads. This represents the case of accessing large contiguous regions of shared memory.

Figure 2 plots the overhead in terms of "Milliseconds/MB" for both software OMP implementations while using a 4MB shared array together with the three different chunk sizes. As expected, both implementations give their worst performances when the chunk size is 4 bytes. However, these scale very differently. The SCLIB DSM is home-based and so maintains a master copy of each page.

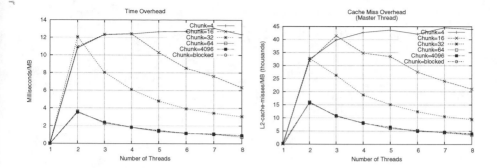

Fig. 3. MCBENCH results on the Sun V1280 SMP. A 4MB array is used with chunk sizes varied from 4 bytes to blocked (4MB/n). The left shows the overhead in terms of Microseconds/MB and the right reports the L2-cache-misses/MB observed by the master thread.

During each iteration of the worst case access pattern, each thread will fetch, twin, diff and later send diffs for $\frac{n-1}{n}$ of the pages in the shared array. This explains the $O(\frac{n-1}{n})$ scaling observed in Figure 2 for both clusters.

In contrast, CLOMP is not home-based. Instead of sending modifications to a central master copy, modifications are maintained by the various writers and must be retrieved from each of these if a thread encounters a subsequent page fault for this page. Since every page has been modified by every thread in the worst case scenario, every page fault during the Read Phase of the benchmark inevitably results in communication with all the other threads. This results in the observed $O(n)$ complexity.

For the 4096-byte and *blocked* chunk sizes, the number of pages accessed for modification or reading can be approximated as $\frac{1}{n} \times \frac{a}{pagesize}$. Thus $O(\frac{1}{n})$ complexity is observed for the CLOMP implementation. For Omni/SCLIB, only $\frac{n-1}{n}$ of the pages accessed are *remote* (i.e. the master copy is at another thread). Due to this, only $\frac{n-1}{n} \times \frac{1}{n}$ of the pages need to be fetched. Therefore the implementation scales at $O(\frac{n-1}{n^2})$. Although, for large enough n this complexity approaches $O(\frac{1}{n})$.

4.2   Analysis: Hardware OMP Implementation

The L1 data cache on the Sun V1280 has a line size of 32 bytes, with 16-byte sub-blocks. The L2 cache is of size 8MB, with 512-byte lines with 64-byte sub-blocks (these form the unit of cache coherency), and is 2-way associative with a random replacement policy.

Figure 3 shows the effect of varying the chunk size c for $a = 4$MB, i.e. over data that can fit entirely in the L2 cache. In addition to the time overhead metric, MCBENCH was instrumented to count the L2 cache misses observed by the master thread. As expected, that there is a reasonably high correlation between the time-based and counter-based overheads. Again, $c = 4$ represents a worst-case

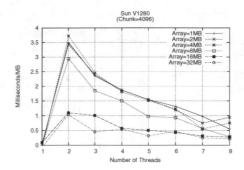

Fig. 4. MCBENCH results on the Sun V1280 SMP. The effect of varying the array size for 4096-byte chunks is shown.

scenario, showing roughly $O(\frac{n-1}{n})$ scalability due to cache-line transfers (including a L1 cache invalidation) that could possibly occur for every element updated. The observed results suggest that the coherency-related hardware has been saturated in this manner. The performance is markedly improved and the scaling becomes roughly $O(\frac{1}{n})$ when the chunk size, c, is increased to 16 and 32 bytes. Finally, when c reaches or exceeds the size of the unit of cache coherency, transfers occur once for every 64-byte region accessed that needs to be made consistent. The backplane fat-tree topology hardware is able to parallelize the coherency traffic effectively at this point, resulting in greatly reduced overhead and a clear $O(\frac{1}{n})$ scaling.

Figure 4 shows the effect of increasing the array size from 1 MB to 32 MB. The results indicate that the consistency overhead is apparently reduced when the array size exceeds the L2 cache capacity. While this may seem at first counterintuitive, at this point we see the distinction between memory and cache consistency overheads emerging, and it must be noted that MCBENCH results are the difference between the access patterns for shared and private arrays. For the latter, a simple cache simulation program showed that for this L2 cache, there is a cache hit rate of only 15% ($a = 16$ MB) and 1% ($a = 32$ MB) upon the first byte in each line that is accessed. Hence, the effect is due to the introduction of memory access overheads for the private array, which, while not as great as the cache coherency overheads for the shared array, reduce some of the difference between the two cases.

5   Conclusions

The overhead reported by MCBENCH is determined by the difference in performance between memory operations on shared and private data. This is achieved by using an access pattern that is designed to force the memory subsystem to perform memory consistency work when shared data is used.

Although MCBENCH was designed in the context of cluster OpenMP implementations, it has been demonstrated that the benchmark can be used on both software and hardware OMP implementations. The results show that the magnitude of the memory consistency overhead is greatly influenced by the structure of the underlying memory subsystem. In particular, varying the chunk size shows that the granularity of coherency is able to sway results from one end of the scale to the other. This is true for both software and hardware based OMP implementations. Beyond the coherency granularity, the same scaling is observed for all three implementations.

Finally, it is important to remember that micro-benchmarks measure specific system overheads by using small well defined operations to exercise that aspect of interest. Thus, it is rare for a single micro-benchmark to explain the performance of a full application completely. Rather, they give an appreciation of the overheads of specific conditions that may manifest during the execution of the application of interest.

Acknowledgments

We are grateful to the APAC National Facility for access to the InfiniBand cluster and to R. Humble and J. Antony for their assistance. This work is part funded by Australian Research Council Grant LP0669726 and Intel Corporation, and was supported by the APAC National Facility at the ANU.

References

1. Adve, S.V., Gharachorloo, K.: Shared memory consistency models: A tutorial. IEEE Computer 29(12), 66–76 (1996)
2. Basumallik, A., Eigenmann, R.: Towards automatic translation of OpenMP to MPI. In: Arvind, Rudolph, L. (eds.) Proceedings of the 19th Annual International Conference on Supercomputing (19th ICS 2005), Cambridge, Massachusetts, USA, jun 2005, pp. 189–198. ACM Press, New York (2005)
3. Bull, J.M.: Measuring synchronisation and scheduling overheads in OpenMP. In: Proc. of the European Workshop on OpenMP (EWOMP 1999) (1999), citeseer.ist.psu.edu/bull99measuring.html
4. Bull, J.M., O'Neill, D.: A microbenchmark suite for OpenMP 2.0. ACM SIGARCH Computer Architecture News 29(5), 41–48 (2001)
5. Harada, H., Tezuka, H., Hori, A., Sumimoto, S., Takahashi, T., Ishikawa, Y.: SCASH: Software DSM using high performance network on commodity hardware and software. In: Eighth Workshop on Scalable Shared-memory Multiprocessors, May 1999, pp. 26–27. ACM Press, New York (1999)
6. Hoeflinger, J.P.: Extending OpenMP to clusters. White Paper Intel Corporation (2006)
7. Huang, L., Chapman, B.M., Liu, Z.: Towards a more efficient implementation of OpenMP for clusters via translation to global arrays. Parallel Computing 31(10-12), 1114–1139 (2005)
8. Karlsson, S., Lee, S.-W., Brorsson, M.: A fully compliant OpenMP implementation on software distributed shared memory. In: Sahni, S.K., Prasanna, V.K., Shukla, U. (eds.) HiPC 2002. LNCS, vol. 2552, pp. 195–208. Springer, Heidelberg (2002)

9. Kee, Y.-S., Kim, J.-S., Ha, S.: ParADE: An OpenMP programming environment for SMP cluster systems. In: Supercomputing 2003, p. 6. ACM Press, New York (2003)
10. Keleher, P., Cox, A., Dwarkadas, S., Zwaenepoel, W.: TreadMarks: Distributed memory on standard workstations and operating systems. In: Proceedings of the 1994 Winter Usenix Conference, pp. 115–131 (1994)
11. Kusano, K., Satoh, S., Sato, M.: Performance Evaluation of the Omni OpenMP Compiler. In: Valero, M., Joe, K., Kitsuregawa, M., Tanaka, H. (eds.) ISHPC 2000. LNCS, vol. 1940, pp. 403–414. Springer, Heidelberg (2000)
12. Müller, M.S.: A Shared Memory Benchmark in OpenMP. In: Zima, H.P., Joe, K., Sato, M., Seo, Y., Shimasaki, M. (eds.) ISHPC 2002. LNCS, vol. 2327, pp. 380–389. Springer, Heidelberg (2002)
13. OpenMP Architecture Review Board. OpenMP Application Program Interface Version 2.5 (May 2005)
14. Sato, M., Harada, H., Hasegawa, A.: Cluster-enabled OpenMP: An OpenMP compiler for the SCASH software distributed shared memory system. Scientific Programming 9(2-3), 123–130 (2001)
15. Sato, M., Harada, H., Ishikawa, Y.: OpenMP compiler for software distributed shared memory system SCASH. In: Workshop on Workshop on OpenMP Applications and Tool (WOMPAT 2000), San Diego (2000)
16. Sato, M., Kusano, K., Satoh, S.: OpenMP benchmark using PARKBENCH. In: Proc. of 2nd European Workshop on OpenMP, Edinburgh, U.K (September 2000), citeseer.ist.psu.edu/article/sato00openmp.html
17. Tezuka, H., Hori, A., Ishikawa, Y.: PM: A high-performance communication library for multi-user parallel environments. Tech. Rpt. TR-96015, RWC (November 1996)
18. Wong, H.J., Rendell, A.P.: The design of MPI based distributed shared memory systems to support OpenMP on clusters. In: Proceedings of IEEE Cluster 2007 (September 2007)
19. Zeffer, H., Hagersten, E.: A case for low-complexity MP architectures. In: Proceedings of Supercomputing 2007 (2007)

Incorporation of OpenMP Memory Consistency into Conventional Dataflow Analysis

Ayon Basumallik and Rudolf Eigenmann

School of Electrical and Computer Engineering
Purdue University, West Lafayette, IN 47907-1285
http://www.ece.purdue.edu/ParaMount

Abstract. Current OpenMP compilers are often limited in their analysis and optimization of OpenMP programs by the challenge of incorporating OpenMP memory consistency semantics into conventional data flow algorithms. An important reason for this is that data flow analysis within current compilers traverse the program's control-flow graph (CFG) and the CFG does not accurately model the memory consistency specifications of OpenMP. In this paper, we present techniques to incorporate memory consistency semantics into conventional dataflow analysis by transforming the program's CFG into an OpenMP Producer-Consumer Flow Graph (PCFG), where a path exists from writes to reads of shared data if and only if a dependence is implied by the OpenMP memory consistency model. We present algorithms for these transformations, prove the correctness of these algorithms and discuss a case where this transformation is used.

1   Introduction

OpenMP [1] has established itself as an important method and language extension for programming shared-memory parallel computers. With multi-core architectures becoming the commodity computing elements of the day, OpenMP programming promises to be a dominant mainstream computing paradigm.

OpenMP is supported by several vendors by means of compilers and runtime libraries that convert OpenMP programs to multi-threaded code. However, current compilers are often limited in the extent to which they use the memory consistency semantics of OpenMP to optimize OpenMP programs. A reason for this is that most data flow analysis employed by state-of-the-art optimizing compilers are based on the traversal of a conventional control-flow graph – a program representation for sequential programs. In sequential programs, data always flows as expressed by the control-flow graph (CFG) and data flow algorithms infer dependences by traversing this CFG. In parallel programs, this is no longer accurate, as data can flow into a thread potentially at any read from a shared variable. To understand such flow, the specific memory consistency model of the employed parallel programming paradigm must be considered.

In this paper, we present techniques to incorporate OpenMP memory consistency semantics into conventional control-flow graphs. Our proposed techniques

R. Eigenmann and B.R. de Supinski (Eds.): IWOMP 2008, LNCS 5004, pp. 71–82, 2008.

transform a conventional control-flow graph (CFG) into an "OpenMP Producer-Consumer Flow Graph" (PCFG), which resembles a conventional CFG and incorporates OpenMP memory consistency semantics into its structure. This enables the use of conventional data flow algorithms for OpenMP programs.

Related approaches to internally representing OpenMP programs for compiler analysis have proposed techniques to incorporate OpenMP control-flow semantics into a program's control-flow graph [2,3]. The present paper is meant to complement, rather than compete with, these related approaches. The focus of this paper is more specifically on techniques to incorporate the memory consistency semantics of OpenMP programs into the internal representation. We shall illustrate why simply incorporating OpenMP control-flow information into the CFG may not be sufficient to account for the effects of the OpenMP memory consistency model. We shall then present formal algorithms to transform a conventional CFG into a representation that accurately reflects OpenMP memory consistency.

The rest of the paper is organized as follows. Section 2 describes the OpenMP Memory Model and introduces transformations that can be applied to a program's CFG to incorporate OpenMP semantics. Section 3 presents algorithms to accomplish the transformations required to create the PCFG, presents a formal proof of correctness of these algorithms and discusses an application of the PCFG. Section 4 discusses related work. Section 5 concludes the paper.

2   The OpenMP Memory Consistency Model

The OpenMP memory consistency model is roughly equivalent to *Weak Consistency* [4]. Writes to shared data by one thread are not guaranteed to be visible to another thread till a synchronization point is reached. OpenMP has both implicit and explicit memory synchronization points. Examples of explicit synchronization points include *barrier* and *flush* directives. Implicitly, there are memory synchronization points at the end of work sharing constructs (unless they have explicit *nowait* clauses) and at the end of synchronization directives like *master* and *critical*. This means, for example, that writes to shared data in one iteration of an OpenMP *for* loop by one thread are not guaranteed to be visible to another thread executing a different iteration of the same loop till the implicit synchronization at the end of the loop is reached.

Figure 1 illustrates some ramifications of how the OpenMP consistency model affects analysis that are based on the program's control-flow graph. The *nowait* clause in loop $L1$ denotes that writes to the array A by one thread in loop $L1$ are not guaranteed to be visible to another thread executing iterations of loop $L2$. However, any analysis based on the corresponding control-flow graph (which incorporates OpenMP control information) shown in the figure will find a path from vertex $v1$ to $v2$ and incorrectly infer that there is a dependence between the write to A in $V1$ and the read of A in $v2$.

On the other hand, the *flush* and *atomic* directives denote that the atomic update to the scalar $tflag$ after loop $L2$ by one thread may be visible to another thread reading $tflag$ in loop $L1$. However, in the graph there is no path from

$v3$ to $v1$ and data flow analysis based on this graph will infer that there is no dependence between the two.

To correctly model these two cases, the control-flow graph needs to be adjusted so that there is a path from the write to a shared variable to a read if the write by one thread is visible to the read by another thread as per OpenMP specifications. A way of doing this for the graph shown in Figure 1 would be to add the edge $e1$ to account for the flush directives, delete the edge $e2$ to account for the nowait clauses and to add edges $e3$ and $e4$ to keep other paths in the graph unbroken even though the edge $e2$ has been deleted. By doing these edge additions and deletions, we create an *OpenMP producer-consumer flow graph* where there is a path from a write W to a read R in the program if and only if the write W occurring on one thread can be visible to the read R occurring on another thread. In certain cases, like the reads and writes connected by edge $e1$ in Figure 1, the read may be before the write in program order.

The next section of this paper presents formal algorithms to create such a *OpenMP Producer-Consumer Flow Graph*, starting from the sequential

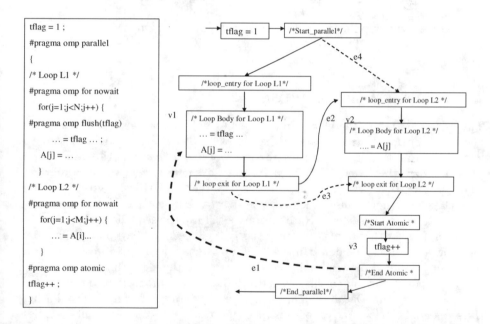

Fig. 1. Incorporation of OpenMP Memory Consistency Semantics: The graph edges drawn with solid arrows are present in the CFG for the program. The dependence implied between writes to array A in loop $L1$ and the read of A in loop $L2$ are relaxed by the *nowait* clause. Therefore, the path from vertex $v1$ to $v2$ in the sequential CFG must be broken to model this relaxation. The combination of *flush* and *atomic* directives imply a dependence between the update to $tflag$ in vertex $v3$ and the read of $tflag$ in vertex $v1$. Therefore, a path must be introduced between $v3$ and $v1$ in the sequential CFG to model this additional dependence. These path adjustments are accomplished by removing edge $e2$ and adding edges $e1, e3, e4$ to the CFG.

control-flow graph for the program. We are incorporating these algorithms into the Cetus [5] infrastructure as part of an OpenMP to MPI translation pass.

3   Incorporation of OpenMP Memory Consistency into the Dataflow Analysis

Our compiler creates an *OpenMP producer-consumer flow graph* using four steps –

1. Identify Shared Data.
2. Incorporate OpenMP control and data synchronization.
3. Relax Sequential Consistency.
4. Adjust for flushes.

We start with the sequential control-flow graph (CFG) for the program. The first step distinguishes between *shared* and *private* data in the program. The second step incorporates OpenMP constructs into the CFG to create an *OpenMP control-flow graph (OpenMP CFG)*. The third and fourth steps first relax and then tighten ordering constraints in the program based on OpenMP memory consistency semantics to transform the OpenMP CFG into an *OpenMP Producer-Consumer Flow Graph (PCFG)*.

3.1   Identification of Shared Data

The very first step in the incorporation of OpenMP semantics is to distinguish between *shared* and *private* data in the program. *Shared data* is defined as data in the program that may be read or written by more than one thread. Private data, on the other hand, is written by only a single thread in the program. The only way that private data is affected by the OpenMP semantics is when it is classified using clauses such as *firstprivate, lastprivate* and *threadprivate*. Thus, dataflow analysis for private data can still use the sequential CFG for the program. It is for the shared data that the PCFG needs to be created before any dataflow analysis can be done. In the very first step, our compiler identifies data that *may* be shared between multiple threads using the algorithm in Figure 2.

At this point, our compiler has separated data in the program into two classes - *shared* and *private*. Dataflow analysis for private data can be now invoked using the sequential CFG for the program. Once that analysis is complete, our compiler transforms the graph to incorporate OpenMP memory consistency semantics prior to invoking dataflow analysis for shared data. For this, our compiler starts by making OpenMP constructs explicit in the CFG to create the OpenMP CFG.

3.2   Making OpenMP Constructs Explicit

Our starting point for this step is the sequential CFG $G = < V, E >$ for the program, where a vertex V represents a basic block in the program and an edge $E = V_1 \rightarrow V_2$ exists if the basic block denoted by V_2 is a successor of

Algorithm *list_shared_variables*
Input : A - An OpenMP program. **Output :** S - A List of *Shared Variables* in A.
Start *list_shared_variables*
1. Set $S = \Phi$
2. **do** $\forall R$, R is an OpenMP *parallel region* in A
3.   Set V = Set of all variables used in R
4.   Set L = Set of all variables declared locally within R
5.   Set PV = Set of all variables explicitly declared *private* for R
6.   Set SV = Set of all variables explicitly declared *shared* for R
7.   Set $S = S \cup (V - L - PV) \cup SV$
8. **end do**
9. **if** $(S = \Phi)$, *exit*, **endif**
10.   **do** $\forall F$, F is *function call* within program A
11.   **do** $\forall Pa$, Pa is a parameter of F
12.   **if** $(Pa \in S)$
13.   Let FP be the *Procedure* that defines F
14.   Let PA be the *Procedure Parameter* of FP
                     corresponding to function parameter Pa
15.   Set $S = S \cup PA$
16.   **end if**
17.   **end do**
18.   **end do**
19.   **if** (Steps 10 through 18 have added new elements to S)
20.   Go to Step 10
21.   **end if**
End *list_shared_variables*

Fig. 2. Algorithm to create list of *Shared Variables* in an OpenMP Program. A key challenge in identifying shared variables is that function calls with shared variables as parameters may introduce additional shared variables that are not explicitly identified as *shared* by OpenMP directives. This algorithm addresses this challenge using the inter-procedural analysis shown in lines 10 through 21.

the basic block denoted by V_1. To incorporate OpenMP constructs into G for the OpenMP program, our compiler inserts vertices corresponding to OpenMP directives. Directives that refer to a set of statements in the program code are represented by *entry* and *exit* vertex pairs. For example, for each OpenMP parallel region in the program, there is a *parallel region entry* and a *parallel region exit* vertex. For each OpenMP critical section in the program, there is a *critical section entry* and a *critical section exit* vertex.

Stand-alone directives such as the *flush* and *barrier* directives are represented with a single vertex in the program flow graph. Each OpenMP *flush* vertex is associated with a *flush set*, which is a list of all shared variables that need to be *flushed* at that point. When the flush set is explicitly specified in the program, the corresponding flush vertex is annotated with this flush set. The *atomic* directive is represented with a pair of *atomic entry* and *atomic exit* vertices around the atomic statement.

Next, our compiler inserts an explicit *barrier* vertex wherever control synchronization is implicit in an OpenMP directive. Thus, *barrier* vertices are added to G at the entry to an exit from parallel regions and at the exit of worksharing regions that do not have *nowait* clauses. *flush* vertices are inserted where a flush is implicit without a barrier, such as at entry to and exit from *critical*, *ordered* and *atomic* regions. Flush sets are derived by the compiler for these inserted *flush* vertices. For example, for *critical* and *atomic* regions, flush sets include the shared variables accessed in these regions. For shared variables that have a *volatile* type, pairs of *flush* vertices enclose every access to these variables.

Thus, at the end of this step, we have an *OpenMP Control Flow Graph* $\hat{G}$ that contains vertices corresponding to OpenMP constructs, *barrier* vertices where control synchronization is implied in the program and *flush* vertices where a data coherence is implied in the OpenMP program.

3.3 Relaxation of Sequential Consistency

With the graph $\hat{G}$ now containing explicit synchronization vertices and vertices corresponding to OpenMP directives, our compiler proceeds to the next step of relaxing sequential consistency constraints using the algorithm *relax_sequential_consistency* shown in Figure 3.

In this algorithm, the compiler deletes edges from the program's control-flow graph, to break paths from writes to subsequent reads of shared data elements where the weak consistency model of OpenMP specifies that the write by one thread may not be visible to the read on another thread. Then the compiler adds edges from the previous synchronization points in the program to preserve paths to the read from writes before the previous synchronization.

At the end of this step, our compiler produces a control-flow graph where any path from a producer to a consumer for a shared variable exists only if this path exists in the original graph and the OpenMP directives in the program do not relax this dependence. In the next step, the compiler adds paths to account for producer-consumer relationships that are additionally introduced by OpenMP directives.

3.4 Adjustment for Flushes

Finally, our compiler uses the algorithm *Adjust_for_Flushes* shown in Figure 4 to adjust for explicit flushes in the program. For line 6 of this algorithm, two flushes are termed *concurrent* in our context if there is no execution order enforced upon them by the program structure. Thus, these may execute in any order, on different threads, between two synchronization points (*barriers*) in the program. To find concurrent flushes, the compiler uses a concurrency analysis for OpenMP [3] which has been used by other researchers as part of static race detection in OpenMP programs.

At this point, the compiler has a control-flow graph that reflects the OpenMP memory consistency model. In this graph, there is a path from a write statement $S1$ to a future read statement $S2$ if and only if the execution of $S1$ by one thread

Algorithm *relax_sequential_consistency*
Input : 1. The OpenMP Control-Flow Graph $\hat{G}$ containing
        explicit synchronization vertices for *barrier* and *flush*
        and *entry* and *exit* vertices for OpenMP directives.
Output : 1. An OpenMP Control-Flow Graph $\hat{G}$ that models
        the Relaxed Memory Consistency of OpenMP.

Start *relax_sequential_consistency*
1. **do** $\forall L$, L is an OpenMP loop,
2.   Remove the back edge from loop entry to loop exit for L.
3. **end do**
4. **do** $\forall V_r$, V_r is an OpenMP *exit* vertex in $\hat{G}$
5.   **if** ($\hat{G}$ contains an edge $V_x \rightarrow V_y$ where
6. V_y is not an OpenMP *barrier* vertex) **then**
7.   Delete edge $V_x \rightarrow V_y$
8.   Let V_{ey} be a *barrier* vertex reachable from V_y without intervening barriers
9.   Let V_dx be a *barrier* vertex that strictly dominates V_x
10.   Add edge $V_x \rightarrow V_{ey}$ to $\hat{G}$
11.   Add edge $V_dx \rightarrow V_y$ to $\hat{G}$
12.   **end if**
11. **end do**
End *relax_sequential_consistency*

Fig. 3. Algorithm to adjust the Control-Flow Graph to remove dependencies according to OpenMP's Memory Consistency specifications

produces an update to memory that is visible to the execution of $S2$ by another thread, as per OpenMP specifications. We refer to this adjusted control-flow graph as the *OpenMP Producer-Consumer Flow Graph (PCFG)*.

3.5 Proof of Correctness

We now present a formal proof of the correctness of the two algorithms presented above.

Theorem 1. *For an OpenMP Producer-Consumer Flow Graph, a Read statement R is reachable from a Write statement $W \Leftrightarrow$ the execution of W by one thread is guaranteed to be visible to the execution of R by another thread according to OpenMP specifications.*

Proof. We begin by first proving proposition in the forward direction – A *Read* statement R is reachable from a *Write* statement W in the OpenMP PCFG $\Rightarrow$ the execution of W by one thread is guaranteed to be visible to the execution of R by another thread according to OpenMP specifications.
    Consider two cases.

Case 1 – R occurs after W in program order.

Algorithm *Adjust_for_Flushes*
Input : 1. The OpenMP Control-Flow Graph $\hat{G}$ for the OpenMP program
        created by algorithm *relax_sequential_consistency*.
Output : 1. An OpenMP Producer-Consumer Flow Graph $\hat{G}$ for the program.

Start *Adjust_for_Flushes*
1. **do** $\forall V_f$, V_f is a *flush* vertex in $\hat{G}$,
2.   **do** $\forall V_f'$, V_f' is a *flush* vertex in $\hat{G}$, $V_f \neq V_f'$
3.   **if** $(V_f \neq V_f'$ and V_f' is not reachable from $V_f)$ **then**
4.   Let S be the *flush-set* of V_f
5.   Let S' be the *flush-set* of V_f'
6.   **if** $(V_f$ and V_f' can be *concurrent* [3]
7.   and $S \cap S' \neq \Phi$) **then**
8.   Add edge $V_f \rightarrow V_f'$ to $\hat{G}$
9.   **end if**
10.   **end if**
11.   **end do**
12. **end do**
End *Adjust_for_Flushes*

Fig. 4. Algorithm to adjust the Control-Flow Graph to incorporate dependencies created by explicit *flushes*

In this case, there can be three scenarios – (i) both R and W are in serial regions, (ii) either R or W is in a serial region and (iii) both R and W are in parallel regions.

If R and W are both in serial sections and there is a path from W to R, then the statement is trivially true.

If W is in a serial region and R is in a parallel region, let E_R be the entry vertex for the parallel region that R is in. E_R dominates R and so any path from W to R must contain E_R. Since there is an implicit synchronization at the beginning and end of each parallel region, E_R is dominated by a *barrier* vertex which must also be in the path from W to R and thus, the execution of W will be visible to an execution of R on any thread. Similarly, if W is in a parallel region and R is in a later serial region, the barrier at the end of this parallel region must be in the path from W to R and thus the execution of W on any thread will be visible to the thread executing the serial region that contains R. If W and R are both in a parallel regions then let E_W be the exit vertex for the OpenMP construct that W is within. Since Algorithm 3 ensures that the successor of E_W is always a *barrier* vertex, there is always a *barrier* vertex in the path from W to R and thus the execution of W will be visible to an execution of R on any thread.

Case 2 – *R* occurs before *W* in program order.
In this case, the path from W to R must contain an edge not present in the original control-flow graph of the program. Additional edges are introduced by line 8 in algorithm *Adjust_for_Flushes* in Figure 4 and by lines 10-11 in algorithm

relax_sequential_consistency in Figure 3. Since these edges contain vertices which are either barriers or flush pairs, the execution of W must be visible to an execution of R on any thread.

We now prove the proposition in the reverse direction – The execution of a *Write* statement *W* by one thread is guaranteed to be visible to the execution of a *Read* statement *R* by another thread according to OpenMP specifications ⇒ R is reachable from W in the OpenMP PCFG.

If W is visible to R on all threads as per OpenMP specifications, then one of two cases must be true –

Case 1 – R is reachable from W in the original program flow graph.

In this case, there must be an intervening *barrier* vertex in the path from W to R since W is guaranteed to be complete before R is started on any thread. We call this *barrier* vertex V_b. The only transformation that deletes edges from the original control-flow graph is line 7 in algorithm *relax_sequential_consistency* in Figure 3. However, the additional edges introduced in lines 10 and 11 of this algorithm ensure that paths from OpenMP *entry* and *exit* vertices to preceding and succeeding barriers are not broken. Thus, a path from W to V_b and from V_b to R is unbroken by the two algorithms. Thus, R is reachable from W in the OpenMP PCFG.

Case 2 – R is not reachable from W in the original program flow graph.

In this case, W must become visible to R because of OpenMP *flush* directives. Thus, there must be a *flush* in the program after W that is reachable from W in the OpenMP CFG. There must also be a *flush* in the OpenMP CFG from which R is reachable. Additionally, these two flushes must be *concurrent*. However, if these flushes are concurrent, then line 8 in algorithm *Adjust_for_Flushes* in Figure 4 will create an edge between them. Thus, R will be reachable from W in the OpenMP PCFG.

Thus, by combining the two propositions proved above, we get "A *Read* statement *R* is reachable from a *Write* statement *W* in the OpenMP PCFG ⇔ the execution of *W* by one thread is guaranteed to be visible to the execution of *R* by another thread according to OpenMP specifications."

3.6   Applications of the OpenMP Producer-Consumer Flow Graph

The OpenMP producer-consumer flow graph is just like a conventional control-flow graph except that it accurately represents producer-consumer relationships between writes to and reads of shared data. Thus, this graph can now form the basis for subsequent dataflow analysis passes for shared variables.

Consider, for example, a typical dataflow analysis pass to find *reaching definitions*. Consider again the program snippet shown in Figure 1. Let $B1$ be the basic block that contains the statement accessing the shared variable $tflag$ in loop $L1$ and let $B2$ be the basic block containing the statement $tflag + +$ after loop $L2$. The algorithm *Adjust_for_Flushes* shown in Figure 4 creates a path from $B2$ to $B1$ in the graph and thus, the definition of $tflag$ in $B2$ would get included in the list of *reaching definitions* for the use of $tflag$ in $B1$.

In our compiler, the OpenMP producer-consumer graph is a key element in a pass to transform OpenMP programs directly to MPI programs [6]. A summary

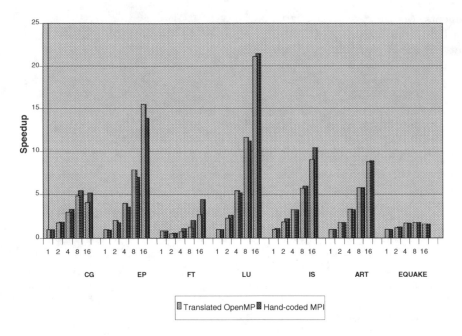

Fig. 5. Performance of seven representative OpenMP applications translated to MPI, compared with their hand-coded MPI counterparts on 16 WinterHawk nodes of an IBM SP2 system. A key step in the OpenMP to MPI translation is the creation of the OpenMP Producer-Consumer Flow Graph.

of the performance of this transformation is shown in Figure 5. This pass performs a whole program analysis of accesses to shared variables and communicates shared data, as it is produced, to all *potential* future consumers. This analysis needs to be conservative. Therefore, accurately representing the relaxation of constraints implied by weak consistency and OpenMP clauses such as *nowait* enables performance optimizations by eliminating certain producer consumer relationships. On the other hand, it includes additional constraints introduced by *flush* operations into the dataflow analysis framework, thereby preserving the correctness of the derived producer-consumer relationships.

4   Related Work

Previous research into the compiler analysis for programs with relaxed consistency models have focused on *delay set analysis* [7,8,9] to ensure correct execution of programs. Others have proposed techniques to use the compiler to hide or abstract the effects of the memory model from the programmer [10,11] and in doing so have relied on specialized graph representations such as the *Concurrent Static Single Assignment* form to represent parallel programs. Our techniques do not have to incorporate delay set analysis since the requisite fences or barriers in our program are already present in the form of OpenMP synchronization statements. Also, rather than modifying the dataflow analysis passes in any way, our technique modifies the sequential control-flow graph for the program, adding and deleting edges to incorporate the effects of weak memory consistency and the additional constraints introduced by OpenMP flushes.

Our work complements recent research to analyze the synchronization structure of OpenMP programs [3]. To build the PCFG our algorithm starts from the sequential CFG. A first addition is to incorporate control-flow semantics for the parallel program to create the OpenMP CFG. This extension has been addressed in recent work [2,3]. In addition, we consider data flow that results from cross-thread communication at *flush* operations and dependences excluded by *nowait* type directives using the techniques proposed in this paper. Recent work on data flow analysis for OpenMP programs [12] has proposed the incorporation of OpenMP semantics by creating *Super Nodes* and *Composite Nodes* in the control flow graph to encapsulate OpenMP constructs and then use different data flow equations for these to incorporate OpenMP semantics. In this paper, we present a different approach that adjusts predecessor and successor relationships between basic blocks to incorporate OpenMP semantics and thus, we do not need to introduce any special data flow equations into the framework.

In our work, we use a simple conservative approach for differentiating between shared and private data. Recent work on autoscoping of data in OpenMP programs [13] proposes alternative approaches to accomplish this. Our work has also benefited from recent efforts to further elucidate the OpenMP memory consistency specifications [14] and to formalize the OpenMP memory model [15].

5   Conclusions

In this paper, we have presented techniques for incorporating OpenMP memory consistency semantics into conventional dataflow analysis. Instead of modifying dataflow analysis in any way, our method is to distinguish between shared and private data in the program, use the sequential control-flow graph of the program to perform dataflow analysis for private data, then transform the control-flow graph into an OpenMP producer-consumer flow graph that reflects the effects of OpenMP memory consistency and to use this graph to perform data flow analysis for shared data.

Our transformation has three essential steps - (*i*) distinguishing shared and private data, (*ii*) incorporating relaxed consistency semantics into the control-flow graph and (*iii*) incorporating any additional constraints introduced by the programming model (by *flush* operations). Thus, our transformations are broadly applicable for any parallel programming paradigm whose memory consistency model can be specified by (*i*) how it differentiates shared and private data, (*ii*) ordering constraints that are relaxed and (*iii*) additional ordering constraints that are introduced.

We use the techniques presented in this paper in the Cetus compiler as part of a set of transformations to translate OpenMP programs to MPI programs. These techniques are essential both for preserving the correctness of the translation and for performance optimization. We believe that these techniques hold promise in a broad spectrum of transformations for a variety of parallel programming models when the memory consistency semantics differ from sequential consistency.

References

1. OpenMP Forum. OpenMP: A Proposed Industry Standard API for Shared Memory Programming. Technical report (October 1997)
2. Satoh, S., Kusano, K., Sato, M.: Compiler Optimization Techniques for OpenMP Programs. In: Proc. of the Second European Workshop on OpenMP (EWOMP 2000) (September 2000)
3. Lin, Y.: Static Nonconcurrency Analysis of OpenMP Programs. In: Proceedings of the first International Workshop on OpenMP (IWOMP 2005) (2005)
4. Adve, S.V., Hill, M.D.: A Unified Formalization of Four Shared-Memory Models. IEEE Trans. on Parallel and Distributed Systems 4(6), 613–624 (1993)
5. Lee, S.-I., Johnson, T.A., Eigenmann, R.: Cetus - An Extensible Compiler Infrastructure for Source-to-Source Transformation. In: Rauchwerger, L. (ed.) LCPC 2003. LNCS, vol. 2958, pp. 539–553. Springer, Heidelberg (2004)
6. Basumallik, A., Eigenmann, R.: Towards automatic translation of openmp to mpi. In: ICS 2005: Proceedings of the 19th annual International Conference on Supercomputing, Cambridge, Massachusetts, USA, pp. 189–198. ACM Press, New York (2005)
7. Shasha, D., Snir, M.: Efficient and correct execution of parallel programs that share memory. ACM Trans. Program. Lang. Syst. 10(2), 282–312 (1988)
8. Krishnamurthy, A., Yelick, K.: Analyses and optimizations for shared address space programs. Journal of Parallel and Distributed Computing 38(2), 130–144 (1996)
9. Fang, X., Lee, J., Midkiff, S.P.: Automatic fence insertion for shared memory multiprocessing. In: ICS 2003: Proceedings of the 17th annual international conference on Supercomputing, pp. 285–294. ACM Press, New York (2003)
10. Lee, J., Padua, D.A.: Hiding relaxed memory consistency with a compiler. IEEE Trans. Comput. 50(8), 824–833 (2001)
11. Midkiff, S.P., Lee, J., Padua, D.A.: A compiler for multiple memory models. Concurrency and Computation: Practice and Experience 16, 197–220 (2004)
12. Huang, L., Sethuraman, G., Chapman, B.: Parallel Dataflow Analysis for OpenMP Programs. In: Proceedings of the International Workshop on OpenMP (IWOMP 2007) (June 2007)
13. Lin, Y., Terboven, C., an Mey, D., Copty, N.: Automatic Scoping of Variables in Parallel Regions of an OpenMP Program. In: Chapman, B.M. (ed.) WOMPAT 2004. LNCS, vol. 3349, pp. 83–97. Springer, Heidelberg (2005)
14. Hoeflinger, J., de Supinski, B.: The OpenMP Memory Model. In: Proceedings of the first International Workshop on OpenMP (IWOMP 2005) (2005)
15. Bronevetsky, G., de Supinski, B.: Complete Formal Specification of the OpenMP Memory Model. In: Proceedings of the second International Workshop on OpenMP (IWOMP 2006) (2006)

STEP: A Distributed OpenMP for Coarse-Grain Parallelism Tool

Daniel Millot, Alain Muller, Christian Parrot,
and Frédérique Silber-Chaussumier

GET/INT, Institut National des Télécommunications, France

Abstract. To benefit from distributed architectures, many applications need a coarse grain parallelisation of their programs. In order to help a non-expert parallel programmer to take advantage of this possibility, we have carried out a tool called STEP (*Système de Transformation pour l'Exécution Parallèle*). From a code decorated with OpenMP directives, this source-to-source transformation tool produces another code based on the message-passing programming model automatically. Thus, the programs of the legacy application can easily and reliably evolve without the burden of restructuring the code so as to insert calls to message passing API primitives. This tool deals with difficulties inherent in coarse grain parallelisation such as inter-procedural analyses and irregular code.

1 Introduction

On the one hand, parallel applications must evolve with hardware architectures ranging from computational grids to multi-core/many-core architectures. Indeed, since parallel programming seeks performance, parallel algorithms will be different depending on architectures in order to take advantage of their specificities. On the other hand, when developing a parallel version of some sequential application (or a new parallel application from scratch), parallel programmers have currently the choice between two alternatives to program their application: either MPI for message-passing programming [PIF95] or OpenMP for shared-memory programming [arb04]. Both ways have advantages and drawbacks. MPI provides a way for the programmer to control the behaviour of his parallel application precisely and especially the time spent in communication or computation. For this reason, MPI programming is considered low-level, and using it for development is time consuming. Furthermore the SPMD programming model of MPI provides a fragmented view of the program [CCZ06]. MPI code is intermingled with the original sequential code and this makes the global structure of the resulting code difficult to read. In contrast programming with OpenMP directives provides a simple way to specify which parts should be executed in parallel and helps keeping a "global view" of the original application. Driven by the directives, the actual parallelisation is delegated to a compiler. The main drawback of OpenMP is to be restricted mainly to shared-memory architectures. Furthermore hybrid programming OpenMP/MPI is necessary to exploit hybrid architectures such as clusters of many-cores.

R. Eigenmann and B.R. de Supinski (Eds.): IWOMP 2008, LNCS 5004, pp. 83–99, 2008.

Considering this dilemma, we propose an approach which tries to make the best of both ways : keeping the relative simplicity of programming with OpenMP directives and developing MPI applications ready to be tuned to run efficiently on distributed memory architectures. Our goal is to propose a MPI implementation of OpenMP directives specifically well-suited to express coarse-grain parallelism. The program written by the programmer remains generic in the sense that it can be either directly compiled by an openMP compiler to be run on some shared-memory architecture, or transformed into an MPI source code that can be further optimised by an expert in order to make the best of non shared-memory architectures such as clusters. In this paper, we propose a tool called STEP (*Système de Transformation pour l'Exécution Parallèle*) that, being based on the source-to-source transformation PIPS workbench [AAC+94], semi-automatically generates MPI programs for some widely-used coarse-grain parallel application structures.

The paper is organised as follows : the next section describes the context and gives a typical use case to emphasize the motivations of our work. In section 3 we describe the prototype we have developed and the results. In section 4, we describe the related work before conclusion.

2   Context

To benefit from distributed architectures, many applications need a coarse grain parallelisation of their programs. Therefore the goal of our work is two-fold:

- find a way to express the potential parallelism of an application that can be derived into both shared and distributed memory parallelisation;
- find a way to parallelise legacy applications without preventing evolutions.

2.1   The Goals of Our Work

Expressing parallelism within an application without corrupting it is the main advantage of the OpenMP paradigm. Using techniques developed in the field of automatic parallelisation, we conceived STEP, a tool that transforms a program semi-automatically into an MPI source program and produces MPI code close to hand-written MPI code. The programmer adds OpenMP directives and then the tool generates a MPI program automatically. The collaboration between the user and the tool provides a semi-automatic parallelisation.

Semi-automatic parallelisation. One of our major goals in this work is to determine to what extent a compiler can help the programmer in developing MPI programs. On the one hand, coarse-grain parallelism is difficult to extract automatically. Automatic parallelism is able to extract fine-grain parallelism automatically. However these techniques could never be used to extract coarser-grain parallelism. There are multiple reasons for this as for example the coarser the grain is, the stronger the analyses must be dealing with inter-procedural analyses, irregular code... Furthermore when dealing with more statements, parallelism may logically exist but can be prevented because of the way the program is written: false dependency because of the use of variables... On the other hand, programmers of the legacy applications have a logical view of the global structure of their applications and thus can easily add simple directives to specify which tasks could be done in parallel.

Combining intelligence of the programmer with techniques of the automatic parallelisation field gives an opportunity to use powerful program transformation tools to discharge the programmer from the parallelising task.

A subset of the OpenMP standard. Another goal of our work is to facilitate parallel programming especially for the programmer of legacy applications who is not a parallel programming specialist. OpenMP programming is relatively straightforward if we stand to main work sharing directives such as "parallel for" and "sections". Nevertheless taking into account all possible directives, clauses and SPMD programming with runtime routines, OpenMP programming can become as tricky as MPI programming and suffer from the same disadvantages as MPI, that is to say, produce code difficult to read and maintain, and error-prone. Firstly, the idea is to restrict our MPI transformation to coarse-grain parallelism to be able to generate efficient code for distributed-memory architectures. Secondly, some clauses are difficult to use and could be determined by analysing tools. For instance, "private" and "shared" data inside a parallel construct can in some cases be determined from data dependency analyses that could also detect potential problems. Alternatively, when inserted by the programmer, those clauses could be a support to analyses that fail without them.

Generation of source code. Parallel programmers seek performance and for that sake, don't necessarily trust black-boxes. Generating source code provides the user a way to understand the parallel program and furthermore gives him an opportunity to tune the generated code to improve its performance. In addition to that, the generated parallel code should propose a generic structure to allow the user to replace part of the generated code with his own code. Thus one important goal of STEP is to provide source-to-source transformation. This has also the advantage, in the end, that the written code makes no hypothesis about the target architecture.

2.2   An Illustrative Use Case

It is often critical in a collaborative scientific context to easily derive parallel versions of an application which are dedicated to different types of architectures. Here we give the example of a real application which gives a flavour of applications eligible for the STEP processing. It is an application implementing a physical optics algorithm that computes the radiation patterns of non-planar aperture antennas over a range of observation angles [BL05].

The target application is composed of two disctinct phases: *computation*1 and *computation*2. Data dependencies imply that the input of the first computation phase array *array*1 is read entirely by all parallel tasks so it should be shared by all these tasks; the first computation can be processed independently on all parallel tasks; and the output of this first phase, *array2*, can then be partitioned on parallel tasks. This pattern applies to the second computation too: *array2* should globally be shared by all parallel tasks while the output of this second phase, *array3*, can be distributed on parallel tasks.

In the shared-memory execution, the fork-join model is used. Arrays *array1*, *array2* and *array3* are shared. There is no need for any explicit synchronisation since there is no write conflict. A barrier is used to synchronise the two computational phases. In the distributed-memory execution, the SPMD model is used. Array *array1* is duplicated for every process. Array *array2* resulting from the first computation phase is distributed among the processes and then an all-to-all exchange provides necessary data for the second computation phase to all processes. The resulting array *array3* is distributed on all processes and can be gathered on one process if necessary.

Given these execution organisations, corresponding OpenMP and MPI parallel programs can be represented as in listings 1.1 and 1.2 respectively.

Listing 1.1. OpenMP implementation

```
real(N) :: array1, array3
real(N) :: array2

initialisation()

!$OMP PARALLEL DO
do i=1,N
    array2(i)= computation1(array1)
end do
!$OMP END PARALLEL DO

!$OMP PARALLEL DO
do i=1,N
    array3(i)= computation2(array2)
end do
!$OMP END PARALLEL DO
```

Listing 1.2. MPI implementation

```
real(N) :: array1, array3
real(M) :: array2

initialisation()

TASK_PARTITIONING()
DETERMINATION_OF_LOOP_INDICES()

do i=istart,iend
    array2(i)= computation1(array1)
end do

ALL_TO_ALL_DATA_REDISTRIBUTION()
DETERMINATION_OF_LOOP_INDICES()

do i=istart2,iend2
    array3(i)= computation2(array2)
end do

DATA_GATHERING()
```

The OpenMP standard guarantees that:

- each thread will access shared arrays *array1*, *array2* and *array3*;
- *i* is private to a thread;
- there is an implicit data flush and an implicit barrier at the end of the *PARALLEL DO* constructs.

The MPI program is more complicated to develop. Data and tasks must be explicitly partitioned depending on the number of processes and on process identifiers. Temporary buffers might have to be allocated. Communications must be handled and may involve personalised all-to-all communication. Furthermore it makes the code much more intricate and difficult to read. Both OpenMP and MPI versions being necessary, a semi-automatic tool able to derive them from the sequential program is clearly welcome.

3   Description of the STEP Prototype

STEP has been developed in the PIPS workbench detailed below. The PIPS project (Inter-procedural Parallelisation of Scientific Programs) [AAC+94] has been developed at ENSMP/CRI since 1988 and is distributed under the term of the GNU public license. PIPS has been used as support to develop new anal yses and program transformations for Fortran programs by several teams from CEA-DAM, Southampton University, SRU, ENST Bretagne and ENS Cachan. Nowadays PIPS is composed of more than 200 000 lines of C code. Since it was first developed for automatic parallelisation, PIPS provides powerful code transformations and analyses that can be used on real codes: source-to-source code based generation, inter-procedural analyses and the array regions analysis.

Using the PIPS workbench, the STEP tool implements the three following phases:

1. the outlining phase that consists in the restructuration of the sequential program by outlining statements in parallel sections,
2. the analysis phase that computes SEND array regions based on PIPS array regions analyses,
3. and the compilation phase that generates the MPI parallel code.

3.1   Parallel Execution Model

To limit changes in the sequential program, the statements delimited by directives are outlined, communications and work-scheduling take place in a new generated subroutine. Data are allocated on all processes and at the end of parallel constructs, each process communicates to other processes the data which may be used in the future. All processes redundantly execute serial regions and iterations of OpenMP "parallel do" loops are partitioned between processes.

3.2   PIPS Array Regions Analysis

In PIPS, array regions are represented by convex polyhedra [Cre96]. They are used to summarise accesses to array elements. Due to region representation, the analyses are not necessarily exact and some regions can be over-approximated. Four different types of array regions are computed. The $READ$ and $WRITE$ regions represent the effects of statements and procedures on sets of array elements. However, $READ$ and $WRITE$ regions can not represent array data flow and they are not sufficient for advanced optimisations such as array privatisation. IN and OUT regions have been introduced for that purpose. For a block of statements or a procedure, an IN region is the subset of the corresponding $READ$ region containing the array elements that are imported (i.e. read before being written in the block) and an OUT region contains the exported array elements (i.e. elements assigned in the block before being potentially read outside it).

Program example for array regions analysis. The program given in the listing 1.3 is an example of a program $P1$ composed of a DO loop labelled 20 that is parallelised with OpenMP directives. Inside the loop, the computation subroutine $F1$ is called.

Listing 1.3. Fortran program for array regions analysis

```
      PROGRAM P1
      INTEGER I ,N, F1
      PARAMETER (N=10)
      INTEGER T(N,2) ,A(N−1)
       . . .
!$OMP PARALLEL DO
      DO 20 I = 1, N−1
          A( I ) = F1(T,  I )
20    CONTINUE
!$OMP END PARALLEL DO
       . . .
      END

      INTEGER FUNCTION F1(T, J)
      INTEGER N, J
      PARAMETER (N=10)
      INTEGER T(N,2)

      IF (MOD(J,  2).EQ.0) THEN
          F1 = T(J+1,1)−T(J,1)
      ELSE
          F1 = T(J+1,2)−T(J,2)
      ENDIF

      END
```

Array region analysis on this example. Based on the previous example, PIPS performs array regions analysis at every statement level of the abstract syntax tree. For example, a result of such an analysis can be seen in the listing 1.4, line 33 :

```
<T(PHI1,PHI2)-R-EXACT-{PHI2==1, J<=PHI1, PHI1<=1+J}>
```

It means that at the statement level of the assignment (see listing 1.4, line 35) the $READ$ region represented by R of the array T is exactly the sub-array $T(PHI1, PHI2)$ with : $J <= PHI1 <= 1 + J$ and $PHI2 == 1$.

Regions $READ, WRITE, IN$ and OUT (see subsection 3.2) are tagged respectively R, W, IN and OUT by the pretty-printer. An over-approximated array region is tagged MAY; the tag $EXACT$ refers to a non-approximated array-region (under-approximation are not computed and fixed at $\emptyset$). A such approximation is shown in the listing 1.4, line 30 where the two exact regions line 33 and 37 are cumulated at the level of the IF statement line 32. These array region analyses are performed at each statement level for intra-procedural analysis, as well as gathered at function level for inter-procedural analysis (see line 24 and line 13 at the call statement).

A new source file is generated by the PIPS pretty-printer that displays the results of *READ*, *WRITE*, *IN* and *OUT* array regions analyses (listing 1.4).

Listing 1.4. Resulting *READ / WRITE / IN* and *OUT* array regions displayed in the program

```
1          PROGRAM P1
2          INTEGER I ,N,F1
3          PARAMETER (N=10)
4          INTEGER T(N,2) ,A(N−1)
5            . . .
6    C    <A(PHI1)−W−EXACT−{1<=PHI1 , PHI1<=9}>
7    C    <T(PHI1 ,PHI2)−R−MAY−{1<=PHI1 , PHI1<=10, 1<=PHI2 , PHI2<=2}>
8    C    <T(PHI1 ,PHI2)−IN−MAY−{1<=PHI1 , PHI1<=10, 1<=PHI2 , PHI2<=2}>
9    C    <A(PHI1)−OUT−EXACT−{1<=PHI1 , PHI1<=9}>
10   !$OMP PARALLEL DO
11          DO 20 I = 1 , N−1
12   C    <A(PHI1)−W−EXACT−{PHI1==I , 1<=I , I<=9}>
13   C    <T(PHI1 ,PHI2)−R−MAY−{I<=PHI1 , 1<=PHI1 , PHI1<=1+I , PHI1<=10,
14   C      1<=PHI2 , PHI2<=2, 1<=I , I<=9}>
15   C    <T(PHI1 ,PHI2)−IN−MAY−{I<=PHI1 , 1<=PHI1 , PHI1<=1+I , PHI1<=10,
16   C      1<=PHI2 , PHI2<=2, 1<=I , I<=9}>
17   C    <A(PHI1)−OUT−EXACT−{I==PHI1 , 1<=I , I<=9}>
18          A( I ) = F1(T, I)
19   20    CONTINUE
20   !$OMP END PARALLEL DO
21            . . .
22          END
23
24   C    <T(PHI1 ,PHI2)−R−MAY−{J<=PHI1 , PHI1<=1+J , 1<=PHI2 , PHI2<=2}>
25   C    <T(PHI1 ,PHI2)−IN−MAY−{J<=PHI1 , PHI1<=1+J , 1<=PHI2 , PHI2<=2}>
26          INTEGER FUNCTION F1(T,J)
27          INTEGER N, J
28          PARAMETER (N=10)
29          INTEGER T(N,2)
30   C    <T(PHI1 ,PHI2)−R−MAY−{J<=PHI1 , PHI1<=1+J , 1<=PHI2 , PHI2<=2}>
31   C    <T(PHI1 ,PHI2)−IN−MAY−{J<=PHI1 , PHI1<=1+J , 1<=PHI2 , PHI2<=2}>
32          IF (MOD(J , 2).EQ.0) THEN
33   C    <T(PHI1 ,PHI2)−R−EXACT−{PHI2==1, J<=PHI1 , PHI1<=1+J}>
34   C    <T(PHI1 ,PHI2)−IN−EXACT−{PHI2==1, J<=PHI1 , PHI1<=1+J}>
35          F1 = T(J+1,1)−T(J ,1)
36          ELSE
37   C    <T(PHI1 ,PHI2)−R−EXACT−{PHI2==2, J<=PHI1 , PHI1<=1+J}>
38   C    <T(PHI1 ,PHI2)−IN−EXACT−{PHI2==2, J<=PHI1 , PHI1<=1+J}>
39          F1 = T(J+1,2)−T(J ,2)
40          ENDIF
41          END
```

Combining READ, WRITE, IN and OUT array region analyses helps us determine SEND array regions that represent data that need to be exchanged between processes and generate MPI code.

In the next subsections, we describe the three different steps of code transformation to produce MPI parallel code.

3.3   First Step: Outlining

To limit changes in the sequential program, the statements delimited by directives are outlined. This code transformation keeps the semantics of the original program and allows to add parallelism without alteration in the original code of sequential parts. For instance, the loop below labelled 20 in the listing 1.5 is outlined and replaced by a call to a subroutine called $P1\_DO20$ as shown the listing 1.6.

Listing 1.5. Before outlining

```
PROGRAM P1
INTEGER I,N,F1
PARAMETER (N=10)
INTEGER T(N,2),A(N−1)
    ...
!$OMP PARALLEL DO
    DO 20 I = 1, N−1
        A(I) = F1(T, I)
20 CONTINUE
!$OMP END PARALLEL DO
    ...
END
```

Listing 1.6. After outlining

```
PROGRAM P1
INTEGER I,N,F1
PARAMETER (N=10)
INTEGER T(N,2),A(N−1)
    ...
!$OMP PARALLEL DO
    CALL P1_DO20(I, 1, N−1, N, A, T)
!$OMP END PARALLEL DO
    ...
END

SUBROUTINE P1_DO20(I, I_L, I_U, N, A, T)
INTEGER I, I_L, I_U, N, F1
INTEGER A(1:N−1), T(1:N, 1:2)
DO 20 I = I_L, I_U
    A(I) = F1(T, I)
20 CONTINUE
END
```

The parameters of the prototype of the new subroutine are composed of:

- loop parameters : the loop index and the loop bounds (see parameters I, I_L and I_U in listing 1.6),
- and variables, arrays and parameters used in the outlined statements (see parameters N, A and T in listing 1.6).

The variable and array updates between callers and callees are performed by the call by reference used in Fortran.

3.4   Second Step: Analysis

In order to generate calls to MPI primitives, data that must be exchanged are determined using PIPS array regions analysis. We distinguish four different types of array regions:

- send regions represent updated array regions which need to be sent at the end of a parallel section;
- receive regions represent data and array regions which need to be received at the beginning of a parallel section;
- private regions represent variables and arrays that could be privatised;
- used regions represent necessary memory allocation gathering READ and WRITE regions.

In our execution model (see subsection 3.1), we suppose that sequential computations are done redundantly in parallel by each process. Then at the beginning of the first parallel section, each process owns up-to-date data. At the end of the parallel section each process must send its own data to update the data of other processes. Thus in our prototype we only use the SEND regions to determine which sub-arrays must be communicated to other processes.

Computation of the SEND array regions. The SEND array regions for a given procedure represent updated array regions which need to be sent at the end of a parallel section thus it is included into OUT array regions. Nevertheless, due to the PIPS inter-procedural array regions analysis, the *OUT* region of an array *A* that is a parameter of a procedure *P*, refers to the array region of *A* updated (and used after) by all calls to the *P* procedure through the whole program. However the SEND region of the array *A* corresponds to only one call to the *P* procedure. Thus the SEND region of *A* for procedure *P* is the intersection between the OUT region and the WRITE region of *A*:

$$SEND(A) = OUT(A) \cap WRITE(A)$$

Handling region overlap. Moreover, array regions can be over-approximated. That is to say SEND regions of different processes can overlap each other. Thus it prevents us from directly exchanging sub-arrays described by the SEND regions. In case of overlap, we compute at runtime data that should be exchanged. Besides, the results of array region analyses are convex polyhedra. For simplicity sake, in our prototype, we limit ourselves to communicating "rectangular" sub-arrays. Rectangular sub-arrays are represented by polyhedra with at most one dimension of the array appearing in each constraint that delimits a SEND

Listing 1.7. Region analyses performed on the subroutine *P1_DO20* (excerpt)

```
Region write   P1_DO20 : 2
<A(PHI1)−W−EXACT−{I_L<=PHI1, PHI1<=I_U}>
<I−W−EXACT−{}>
Region out    P1_DO20 : 1
<A(PHI1)−W−EXACT−{1<=PHI1, PHI1<=9, I==11, I_L==1, I_U==9, N==10}>
Region Send   P1_DO20 : 1
<A(PHI1)−W−EXACT−{I_L<=PHI1, 1<=PHI1, PHI1<=I_U, PHI1<=9}>
Region interlaced   P1_DO20 : 0
```

region. In case sub-arrays are not given rectangular by array region analysis, we make an over-approximation to come back to a rectangular case. In this case also, we must compute at runtime data that should be updated.

The following listing 1.7 shows the array regions analysis performed on the outlined subroutine $P1\_DO20$.

3.5   Third Step: Compilation

The compilation phase consists in generating the MPI parallel code and inserting communications and work-scheduling.

For a "parallel do" directive, we create a new routine suffixed with $\_MPI$ that replaces the outlined sequential routine. For instance, in the main program, the call $P1\_DO20$ is replaced by a call to $P1\_DO20\_MPI$ (see listing 1.8 line 6 and line 26).

This new procedure is divided into four parts :

- computation of the loop partitioning: the initial loop range is split in different slices for each process;
- computation of the SEND array regions according to the loop partitioning;
- call of the outlined procedure according to the loop partitioning: before the call, each process owns necessary data for the computation of its loop slice. After the call, a process owns only the updated data in its own SEND region;
- communications of the updated data between processes : to deal with the potential overlap of send data, at the first step, we merge all the SEND regions on a master process and in the second step, the master broadcasts the global send region corresponding to the initial loop bounds.

In our generated code (see listing 1.8), calls at the MPI library are encapsulated into our own Fortran functions that deal with send and receive arrays described by our own internal region representation.

3.6   Results

STEP has reached several goals specified in subsection 2.1. We have implemented a source-to-source program transformation tool that semi-automatically generates MPI programs called STEP. Currently, STEP is in its early development stage. Its input (resp. output) files are Fortran 77 programs with OpenMP directives (respectively with MPI primitives) to express the parallelism.

Handling more OpenMP directives. STEP handles "section" and "parallel do" work-sharing directives for which the number of available computation nodes is known at runtime. As previously said, we do not necessarily intend to handle the entire OpenMP standard. Nevertheless, STEP is currently limited and we want to extend it to several widely-used OpenMP directives for instance the `master` directive for I/O management. Furthermore the loops are currently partitioned using static scheduling assigning one loop chunk to each process. This should

Listing 1.8. The generated MPI parallel source file

```
1       PROGRAM P1
2       include "step.h"
3   C       declarations
4       CALL STEP_Init
5       ...
6       CALL P1_DO20_MPI(I , 1, N−1, N, A, T)
7       ...
8       CALL STEP_Finalize
9       END
10
11      SUBROUTINE P1_DO20_MPI(I , I_L , I_U , N, A, T)
12      include "step.h"
13  C       declarations
14
15  C       Loop splitting
16      CALL STEP_SizeRank(STEP_Size , STEP_Rank)
17      CALL STEP_SplitLoop(I_L , I_U , 1, STEP_Size , I_STEP_SLI)
18
19  C       SEND region computing
20      CALL STEP_COMP_SEND(A_STEP_SR , ... )
21
22  C       Where the work is done
23      I_IND = STEP_Rank+1
24      LLOW = I_SLI(LOWER, I_IND)
25      LUPP = I_SLI(UPPER, I_IND)
26      CALL P1_DO20(I_IND , LLOW , LUPP , N, A, T)
27
28  C       SEND regions communications
29      IF (STEP_Rank.EQ.MASTER) THEN
30          CALL STEP_RecvMergeRegion ( ... )
31      ELSE
32          CALL STEP_SendRegion ( ... )
33      ENDIF
34      CALL STEP_BcastRegion ( ... )
35      END
```

also be extended to other types of scheduling proposed by OpenMP. Although only one level of parallelism is currently supported, nested parallelism is planned to deal both coarse and fine grain parallelism.

At last, STEP does not handle explicit OpenMP synchronisation directives; only implicit synchronisations at the end of "parallel sections" and "parallel do" are performed. Although this limits the developer's possibilities, it reduces all the more the synchronisations between processes. Besides limiting the amount of synchronisations potentially provides a good execution performance.

Handling data. On the contrary to the OpenMP standard, data are private by default. It is the analysis phase and the computation of the SEND regions that determine which data are *shared*. These analyses specify which data must be transmitted/shared from one process to another at the end of a parallel region. Currently the same amount of memory is allocated on all processes. This is mainly due to Fortran 77. As this is a important limitation of the STEP prototype, we want as a short-term improvement distribute arrays when analyses are successful, for instance at least for arrays with regular access. Besides, our first STEP prototype converts polyhedric regions into rectangular regions in order to generate simple MPI communication datatypes. This could be improved to benefit from the array region representation in PIPS [Cre96].

Communication patterns. Currently, the SEND region analysis leads to use different auxiliary functions to deal with data exchanges :

– when the SEND regions are computed without over-approximation, our tool generates exact MPI messages to perform the exchange;
– when the SEND regions are computed with over-approximation and without interlacing, the same function performs the exchange of all the data specified by the SEND region;
– when the SEND regions are over-approximated and interlaced, another function is used to deal with the interlaced access. A runtime solution has been implemented. It consists in a first step in exchanging all data in the SEND region, and in the second step in systematically comparing original and updated values. Nevertheless, this simple solution with an important overhead allows to detect concurrent accesses.

Currently, the all-to-all data redistribution is centralised for simplicity sake. This could be further refined since array region analyses provide all the necessary information.

At last, current analyses need to be improved to deal with reductions: recognising a reduction pattern in the sequential code when no OpenMP reduction clause is present and then generating the appropriate MPI code. In case where a reduction pattern is not recognised, STEP should be able to handle the OpenMP reduction clause.

Quality of the generated source code. At last, the generated code is close to hand-written code. First, the original sequential code is not altered. The sequential parts that are executed in parallel can be found unmodified in outlined functions. Second, code transformations keep track of the original variable names, and proposes readable function names. At last, MPI communications are close to hand-written communications since the volume of the communicated data is given by array region analysis. Nevertheless it is already possible to use the STEP generated code, read it and modify to tune MPI communications if necessary. For this purpose, auxiliary functions provide several services, as for instance index conversion functions to convert array indices.

At last, the STEP tool is currently limited to Fortran 77 but an extension to C is considered.

4   Related Work

"Distributed OpenMP" refers to research projects that help running an OpenMP application on distributed-memory architectures.

Historically, most distributed OpenMP projects addressed Software Distributed Shared Memory (SDSM) architectures. SDSM architectures rely on a software layer in order to manage data placement on the nodes of a distributed-memory architecture and keep memory consistency between nodes. For instance, the OMNI OpenMP compiler [SSKT99] is based on a DSM runtime system inserting check codes before each load/store to/from the shared data space. Those check codes are implemented using a communication library supporting one-sided remote memory transfer and synchronization between nodes. Several implementations of "distributed OpenMP" are based on the SDSM Treadmarks system [HLCZ00, BME02]. Those approaches suffer from very fine-grain communication patterns. In order to avoid unnecessary data checks and group them when possible, those projects have evolved to use compiler analyses intensively in order to generate efficient code. The OMNI compiler uses analyses to eliminate check codes outside parallel regions, eliminate redundant check codes and merge multiple check codes. In order to do so, the *static extent* and the *dynamic extent* are defined. The static extent corresponds to the statements lexically enclosed within an OpenMP construct. The dynamic extent includes the functions called from within the construct. By default, the compiler determines data mapping according to the scheduling of the loop which references the data. One further optimization proposed by the OMNI compiler is to optimize data transfers between two parallel regions. Using Treadmarks, Basumallik et al. [BME02] proposed optimizations such as data prefetch, barrier elimination and data privatization. Barrier elimination and data privatization both rely on strong compiler analyzes. Barrier separating two consecutive loops can be eliminated when permitted by data dependencies. Shared data with read-only access during a section can be privatized by copy-in during this section. Shared data that are exclusively accessed by the same processor can also be privatized. There is also some work related to automatic data placement and adding HPF-like directives to OpenMP to distribute data [MMS00]. The Cluster OpenMP software system proposed by Intel [Hoe06] allows OpenMP programs to run on clusters. It relies also on the DSM Treadmarks system. As can be assumed, it is mainly suitable for applications with small amount of read / write sharable data and few synchronisation. SDSM-based distributed OpenMP projects represent opposite approaches from ours since they start from the finest communication grain and then use program analyses to communicate less. Our tool STEP targets coarse-grain parallelism, thus aims at issuing MPI primitives only at specific points in the program.

Several projects generate hybrid MPI and SDSM programs from OpenMP directives. Based on the OMNI compiler, the PaRADE project [KKH03] replaces synchronization and work-sharing directives associated with small data structures by MPI explicit collective communication. The `critical` synchronization directive is translated in an `MPI_Allreduce` as well as the `atomic` directive. The `single` work-sharing directive is translated in `MPI_Bcast` whereas the `for` directive has no message-passing translation. Generating MPI, the approach also considers fine-grain communication and computation patterns while we think that MPI is best-suited for coarser-grain parallelizations. Based on the Polaris compiler, the OpenMP compiler proposed by Eigenmann *et al.* [EHK+02] generates MPI programs in case of regular data pattern and Treadmarks SDSM programs in irregular cases. Three classes of data are distinguished: *private data*, *distributed data* for simple usage pattern and *shared data* for irregular pattern. Irregular or unknown patterns are handled by the SDSM system. Regular patterns are handled with message-passing primitives (see HPF compiler techniques): first, determine send/receive pairs for data accessed but not owned by threads; then determine the intersection between the region of an array (represented by Linear Memory Access Descriptor (LMAD)) accessed in one thread and the region owned by another thread; at last determine the `overlap` using the LMAD intersection algorithm of Hoeflinger and Paek [PHP98]. Several cases of intersection between array regions accessed by processors are discussed in the paper. For a read reference, more data can be fetched (superset mode). For a write reference, the write-back must be precise otherwise it generates a failure. A non-empty `overlap` indicates necessary communications and therefore proper message-passing calls are generated. In the paper, several OpenMP extensions are mentioned: for data distribution as HPF directives, for explicit communication operations outside parallel regions with the `copy` clause and for computation distribution with the `schedule` clause to specify which thread computes which iteration and the `home` clause to specify the thread owner.

In the OpenMPI project which follows a similar objective to ours, Boku *et al.* [BSMT04] propose a programming tool for OpenMP-like incremental parallelization based on MPI scheme. They introduce specific directives to express parallelism based on domain decomposition. They focus on data and take into account neighbour communications to exchange border elements. So doing, they obtain some dedicated code which lacks genericity. These new directives can be very interesting when program analyses fail to express these specific communication patterns.

The closest approach to ours is proposed by Basumallik and Eigenmann [BE05, BME07]. They propose a source-to-source OpenMP to MPI transformation based on the Cetus infrastructure (which is the next version of Polaris). All participating processes redundantly execute serial regions and parallel regions marked by *omp master*. Iterations of OpenMP *for* loops are statically partitioned. Shared data is allocated on all processes using a producer/consumer paradigm. At the end of a parallel construct, each participating process communicates the shared data it has produced that other processes may use. This approach is based on a strong program analysis tool. As a matter of fact, the compiler constructs a control flow graph (with each vertex corresponding to a program statement) and records array access summaries with Regular Section Descriptors (RSDs) by annotating the vertices of the control flow graph. Havlak and Kennedy [HK91] compare array region representation and in particular RSD and convex regions used in PIPS. Both methods have advantages and drawbacks depending on applications. In both projects array regions are then propagated through the interprocedural analysis. In addition to READ and WRITE array regions, IN and OUT regions are computed in PIPS implying additional analysis.

All in all comparing compilation, transformation and analysis tools as Cetus or PIPS is not straightforward. Nevertheless generated codes by both tools could be compared in the short term.

5   Conclusion and Future Work

Based on a solid parallel programming experience, our goal is to develop a programming environment that is a trade-off between 1) the current situation of the parallel programmer who programs the entire application by himself based on OpenMP and MPI 2) and the compilation community which has developed powerful program analyses and transformations.

The STEP tool is the first phase. Based on the PIPS workbench, STEP generates MPI source code from some OpenMP directives. Thanks to the PIPS workbench, this implementation was relatively straightforward. As a matter of fact, PIPS provided us with both a workbench for program transformation and a clearly defined internal representation as well as powerful inter-procedural array region analyses. This paper presents early achievements that we can conclude with several perspectives.

Short-term perspectives. This STEP tool must be completed with several technical improvements described earlier in subsection 3.6 as OpenMP, communications... The most important one is to really distribute and not allocate entire arrays since this strongly limits the scope of potential parallel applications. Furthermore this work must be completed with tests on popular benchmarks and speedup figures.

Middle-term perspectives. We want to take into account multilevel parallelism and generate hybrid MPI/OpenMP applications from OpenMP programs. Focusing on coarse-grain parallelisation for MPI parallel code generation, we could delegate fine-grain parallelisation to OpenMP directives for shared-memory execution on multi-core/many-core nodes for instance. In this perspective, a program analysis could compute the ratio between computation and communication and propose to distribute the data on different nodes or not.

Long-term perspectives. The STEP generated MPI code is close to hand-written parallel code in specific cases. We want to further focus on the interaction between STEP and the user. Our goal is to build a tool that is able to work in collaboration with the user and for instance ask for more guidance from the user in case analyses for data dependencies fail. In these "failure" cases, OpenMP programming could be refined by the user adding OpenMP clauses such as `private`, `shared`, `reduction`...

To conclude, we believe that this approach is very promising and we intend to improve this work in these three directions.

Acknowledgements

The work reported in this paper was partly supported by the European ITEA2 ParMA (Parallel programming for Multicore Architectures) Project [1].

References

[AAC+94] Ancourt, C., Apvrille, B., Coelho, F., Irigoin, F., Jouvelot, P., Keryell, R.: PIPS — A Workbench for Interprocedural Program Analyses and Parallelization. In: Meeting on data parallel languages and compilers for portable parallel computing (1994)

[arb04]   OpenMP architecture review board. OpenMP Application Program Interface (2004)

[BE05]   Basumallik, A., Eigenmann, R.: Towards Automatic Translation of OpenMP to MPI. In: Proceedings of the 19th ACM International Conference on Supercomputing (ICS) (2005)

[BL05]   Boag, A., Letrou, C.: Multilevel Fast Physical Optics Algorithm for Radiation From Non-Planar Apertures. IEEE Transactions on Antennas and Propagation 53(6) (June 2005)

[BME02]   Basumallik, A., Min, S.-J., Eigenmann, R.: Towards OpenMP Execution on Software Distributed Shared Memory Systems. In: International Workshop on OpenMP: Experiences and Implementations, WOMPEI 2002 (2002)

[BME07]   Basumallik, A., Min, S.-J., Eigenmann, R.: Programming Distributed Memory Systems Using OpenMP. In: 12th International Workshop on High-Level Parallel Programming Models and Supportive Environments (2007)

[BSMT04] Boku, T., Sato, M., Matsubara, M., Takahashi, D.: OpenMPI - OpenMP like tool for easy programming in MPI. In: Sixth European Workshop on OpenMP (2004)

[CCZ06]   Chamberlain, B.L., Callahan, D., Zima, H.P.: Parallel Programmability and the Chapel Language. The International Journal of High Performance Computing Applications, Special Issue on High Productivity,Programming Languages and Models (2006)

[Cre96]   Creusillet, B.: Array region analyses and applications. PhD thesis, École Nationale Supérieure des Mines de Paris (1996)

[EHK+02] Eigenmann, R., Hoeflinger, J., Kuhn, R., Padua, D., Basumallik, A., Min, S.-J., Zhu, J.: Is OpenMP for Grids. In: NSF Next Generation Systems Program Workshop held in conjunction with IPDPS (2002)

[HK91]   Havlak, P., Kennedy, K.: An Implementation of Interprocedural Bounded Regular Section Analysis. IEEE Transactions on Parallel and Distributed Systems 2(3), 350–360 (1991)

[HLCZ00] Hu, Y.C., Lu, H., Cox, A.L., Zwaenepel, W.: OpenMP for Networks of SMPs. Journal of Parallel and Distributed Computing 60(12) (2000)

[Hoe06]   Hoeflinger, J.P.: Extending OpenMP to Clusters. Technical report, Intel Corporation (2006)

[KKH03]   Kee, Y.-S., Kim, J.-S., Ha, S.: ParADE: An OpenMP Programming Environment for SMP Cluster Systems. In: Conference on High Performance Networking and Computing (2003)

---

[1] http://www.parma-itea2.org/

[MMS00]   Merlin, J., Miles, D., Schuster, V.: Distributed OpenMP: Extensions to
           OpenMP for SMP Clusters. In: Second European Workshop on OpenMP,
           EWOMP 2000 (2000)
[PHP98]   Paek, Y., Hoeflinger, J., Padua, D.: Simplication of Array Access Pat-
           terns for Compiler Optimizations. In: Proceedings of ACM SIGPLAN 1998
           (1998)
[PIF95]   Message Passing Interface Forum. MPI: A Message-Passing Interface Stan-
           dard (1995)
[SSKT99]   Sato, M., Satoh, S., Kusano, K., Tanaka, Y.: Design of OpenMP Com-
           piler for an SMP Cluster. In: Proceedings of First European Workshop on
           OpenMP (EWOMP) (1999)

Evaluation of OpenMP Task Scheduling Strategies

Alejandro Duran, Julita Corbalán, and Eduard Ayguadé

Barcelona Supercomputing Center
Departament d'Arquitectura de Computadors
Universitat Politècnica de Catalunya
Jordi Girona, 1-3, Barcelona, Spain
{aduran,juli,eduard}@ac.upc.edu

Abstract. OpenMP is in the process of adding a tasking model that allows the programmer to specify independent units of work, called tasks, but does not specify how the scheduling of these tasks should be done (although it imposes some restrictions). We have evaluated different scheduling strategies (schedulers and cut-offs) with several applications and we found that work-first schedules seem to have the best performance but because of the restrictions that OpenMP imposes a breadth-first scheduler is a better choice to have as a default for an OpenMP runtime.

1 Introduction

OpenMP grew out structured around parallel loops and was meant to handle dense numerical applications. The simplicity of its original interface, the use of a shared memory model, and the fact that the parallelism of a program is expressed in directives that are loosely-coupled to the code, all have helped OpenMP become well-accepted today. However, the sophistication of parallel programmers has grown in the last 10 years since OpenMP was introduced, and the complexity of their applications is increasing. Therefore, the forthcoming OpenMP 3.0[13] adds a new tasking model[2] to address this new programming landscape. The new directives allow the user to identify units of independent work, called `tasks`, leaving the scheduling decisions of how and when to execute them to the runtime system.

In this paper we explore different possibilities about the scheduling of these new `tasks`. We have extended our prototype runtime[15] with two scheduling strategies: a breadth-first approach and a work-first approach. We have also implemented several queueing and work-stealing strategies. Then, we have evaluated combinations of the different scheduling components with a set of applications. We also evaluated how these schedulers behave if the application uses `tied tasks`, which have some scheduling restrictions, or `untied` ones, wich have no scheduling restrictions.

The remaining of the paper is structured as follows: Section 2 describes our motivation and the related work, Section 3 describes the different schedulers we

R. Eigenmann and B.R. de Supinski (Eds.): IWOMP 2008, LNCS 5004, pp. 100–110, 2008.

have implemented, Section 4 shows the evaluation results and finally Section 5 presents the conclusions of this work.

2   Motivation and Related Work

The Intel *work-queueing* model [14] was an early attempt to add dynamic task generation to OpenMP. This proprietary extension to OpenMP allows hierarchical generation of tasks by nesting `taskq` constructs. The NANOS group proposed `dynamic sections` [4] as an extension to the standard `sections` construct to allow dynamic generation of tasks.

Lately, a committee from different institutions developed a task model[2] for the OpenMP language that seems that it will be finally adopted[13]. One of the things this proposal leaves open is the scheduler of tasks that should be used.

Scheduling of tasks is a very well studied field. There are two main scheduler families: those that use breadth-first schedulers (see for example the work from Narlikar[12] and those that use work-first schedulers with work-stealing techniques (see for example Cilk[7] and Acar et al.[1]). Korch et al.[9] made a very good survey of different task pool implementations and scheduling algorithms and evaluated them with a radiosity application. Many of these works have found that work-first schedulers tend to obtain better performance results.

Several works have studied how to reduce the overhead of task creation by means of using cut-off strategies. They have found that strategies based on controlling the recursion level tend to work very well[10,11]. Another proposal, uses the size of data structures[8] to control task creation but it depends on the compiler understanding complex structures like lists, which is difficult in the C or Fortran languages.

But, it is unclear how all these algorithms will map into the new task model as most of the previous work was in the context of recursive algorithms and where there were no scheduling restrictions at all. But the new task model allows not only non-recursive applications but also applications that mix traditional work-sharing regions with the new task model. Our goal in this work is to evaluate previous techniques in the context of OpenMP and to try to find which ones work best in order to help implementors choose appropriate defaults.

3   Task Scheduling

We have extended our research NANOS runtime[15] with two families of schedulers: breadth-first schedulers and work-first schedulers. These schedulers implement the restrictions about scheduling of `tied tasks` (i.e. tied tasks can only be scheduled on the thread to wich they are tied to). Also, we implemented two cut-off strategies: one based on the level of task recursion and another in the total number of existing tasks.

3.1   Breadth-First Scheduling

Breadth-first scheduling (BF) is a naive scheduler in which every task that is created is placed into the team pool and execution of the parent task continues. So, all tasks in the current recursion level are generated before a thread executes tasks from the next level.

Initially, tasks are placed in a team pool and any thread of the team can grab tasks from that pool. When a task is suspended (e.g. because a `taskwait`), if it is a `tied task` it will go to a private pool of tasks of the thread that was executing the tasks. Otherwise (i.e an `untied task`), it will be queued into the team pool.

Threads will always try to schedule first a task from their local pool. If it is empty then they will try to get tasks from the team pool.

We implemented two access policies for the task pools: LIFO (i.e., last queued tasks will be executed first) and FIFO (i.e., oldest queued tasks will be executed firsts).

3.2   Work-First Scheduling

Work-first scheduling (WF) tries to follow the serial execution path hoping that if the sequential algorithm was well designed it will lead to better data locality.

The WF scheduler works as follows: whenever a task is created, the creating task (i.e. the parent task) is suspended and the executing thread switches to the newly created task. When a task is suspended (either because it created an new one or because some synchronization) the task is placed in a per thread local pool. Again, this pool can be accessed in a LIFO or FIFO manner.

When looking for tasks to execute, threads will look on their local pool. If it is empty, they will try to steal work from other threads. In order to minimize contention we used a strategy where a thread traverses all other threads starting by the next thread (i.e. thread 0 starts trying to steal from thread 1, thread 1 from thread 2, ... and thread n from thread 0). When stealing from another thread pool, to comply with OpenMP restrictions, a task that has become tied to a thread cannot be stolen (note that a `tied task` that has not been yet executed can be stolen). The access to the victim's pool can also be LIFO or FIFO.

We also implemented a stealing strategy that first tries to steal the parent task of the current task. If the parent task cannot be stolen (i.e. because is either already running or waiting on some synchronization) then the default stealing mechanism is used.

The Cilk scheduler[7] pertains to this family of schedulers. In particular, it is a work-first scheduler where access to the local pool is LIFO, tries to steal the parent task first and otherwise steals from another thread pool in a FIFO manner.

3.3   Cutting Off

In order to reduce the size of the runtime structures and, also, reduce the overheads associated to creating tasks, the runtime can decide to start executing tasks immediately. This is usually referred as cutting off.

This is particularly important with breadth-first scheduling as it tends to generate a large number of tasks before executing them. In work-first scheduling the number of tasks that exist at a given time is not so large but it may grow over time because of tasks being suspended at synchronization points.

It is important to note that tasks that are executed immediately because of a cut-off policy are different than the ones that get executed immediately with the work-first scheduler. When cutting off, the new task does not go through to the whole creation process and in many aspects forms part of the creating tasks (e.g. cannot be suspended on its own).

We have implemented two simple but effective cut-off policies:

Max number of tasks (max-task). The total number of tasks that can exist in the pool is computed as a factor of the number of OpenMP threads (i.e. $k * num\_threads$). Once this number is reached new tasks are executed immediately. When enough tasks finish, tasks will be put into the task pool again. In our implementation, we use a default value for k of 8.

Max task recursion level (max-level). When a new task is created, if it has more ancestors than a fixed limit l then the new task is executed immediately. Otherwise it can be placed in the task pool. In our implementation, we use a default value for l of 4.

4 Evaluation

4.1 Applications

We have used the following applications (for more information on the parallelization please check our previous work[3]) for the evaluation of the schedulers:

Strassen. Strassen is an algorithm[6]for multiplication of large dense matrices. It uses hierarchical decomposition of a matrix. We used a 1280x1280 matrix for our experiments.

NQueens. This program, which uses a backtracking search algorithm, computes all solutions of the n-queens problem, whose objective is to find a placement for n queens on an n x n chessboard such that none of the queens attacks any other. We used a chessboard of size 14 by 14 in our experiments.

FFT. FFT computes the one-dimensional Fast Fourier Transform of a vector of n complex values using the Cooley-Tukey algorithm[5]. We used a vector with 33554432 complex numbers.

Multisort. Multisort is a variation of the ordinary mergesort, which uses a parallel divide-and-conquer mergesort and a serial quicksort when the array is too small. In our experiments we were sorting a random array of 33554432 integer numbers.

Alignment. This application aligns all protein sequences from an input file against every other sequence and compute the best scorings for each pair by means of a full dynamic programming algorithm. In our experiments we used 100 sequences as input for the algorithm.

SparseLU. The sparseLU kernel computes an LU matrix factorization. The matrix is organized in blocks that may not be allocated. Due to the sparseness of the matrix, a lot of imbalance exists. In our experiments, the matrix had 50 blocks each of 100x100 floats.

In all applications (except *Alignment*) we marked all tasks as *untied* and we removed any kind of manual cut-off that was there from the programmer to leave total freedom to the scheduler. The *Aligment* application makes heavy use of `threadprivate` and, because of that, we could not mark the tasks as `untied`.

4.2   Methodology

We evaluated all the benchmarks on an SGI Altix 4700 with 128 processors, although they were run on a cpuset comprising a subset of the machine to avoid interferences with other running applications.

We compiled all applications with our Mercurium compiler[4] using gcc with option -O3 as the backend. The serial version of the application was compiled with gcc -O3 as well. The speed-ups were computed using the serial execution time as the baseline and using the average execution time of 5 executions.

We have executed all applications with different combinations of schedulers. Table 1 summarizes the different schedules we have used in the evaluation, their properties (see Section 3 for details) and the name we will be using to refer to them in the next sections.

Table 1. Summary of schedules used in the evaluation

Scheduler Name	Scheduler Type	Pool Access	Steal Access	Steal Parent
bff	breadth-first	FIFO	-	-
bfl	breadth-first	LIFO	-	-
wfff	work-first	FIFO	FIFO	No
wffl	work-first	FIFO	LIFO	No
wflf	work-first	LIFO	FIFO	No
wfll	work-first	LIFO	LIFO	No
cilk	work-first	LIFO	FIFO	Yes

For each schedule we have run the applications using no cut-off and then using the cut-offs we had implemented:the *max-task* and the *max-level*.

Then, we wanted to know how the restrictions of `untied tasks` affected the performance that can be obtained with the different schedulers. So, we have also tried for those combinations that were best from each application but with all tasks `tied` (we control this via an environment variable that the runtime checks).

4.3   Results

In this section we present several lessons we have learned about task scheduling from this evaluation. Because of space considerations we are only showing part of the evaluation.

Lesson 1: Cutting Off: Yes, But How?. Figure 1 shows the speed-ups of three of the applications (*Alignment,FFT* and *Strassen*) and different schedulers. For each of them, three versions are shown: one that uses no cutoff, another that uses the max-level cutoff mechanism and the last that uses the max-task mechanism.

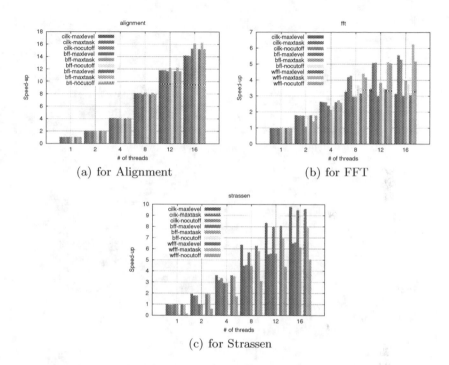

(a) for Alignment

(b) for FFT

(c) for Strassen

Fig. 1. Performance of difference cutoffs

Except for *Alignment*, if a cut-off strategy is not used there is a degradation in the obtained performance. The amount of degradation depends on the scheduler and the application but as a general rule we can see that breath-first schedulers suffer more (see for example *Strassen*) from the lack of a cut-off while work-first schedulers seem to withstand better the lack of a cut-off.

Another observation from these results is that choosing the wrong cut-off can be worse performance-wise than having no cut-off (see for example *FFT* where the max-level cut-off has less speed-up than no cutoff). But, we can also see that for different applications the right cut-off is different (for example compare *FFT* versus *Strassen*).

So, while it seems clear that is important to use a cut-off technique the decision of which to use remains unclear because it depends on the scheduler and also on the exact application.

Lesson 2: Work-First Schedulers Work Best. Figure 2 shows the speed-up obtained with different schedulers (in general we show the most efficient schedulers, but also some others that might be interesting).

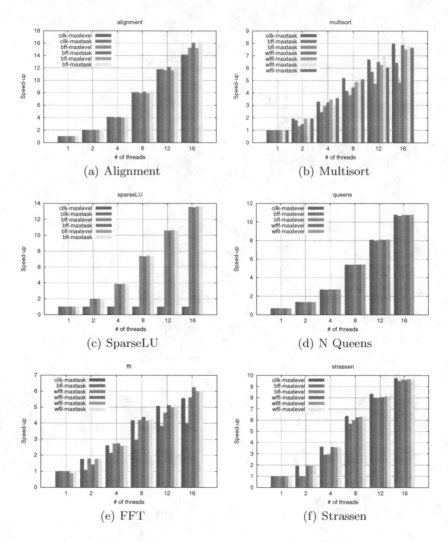

(a) Alignment   (b) Multisort

(c) SparseLU   (d) N Queens

(e) FFT   (f) Strassen

Fig. 2. Speed-ups with different schedulers

We can see that in most applications work-first schedulers get the best speed-up as they tend to exploit data locality better. The exceptions are *Alignment*, where tasks are `tied`, and *SparseLU*, where also a `tied` task limits the performance that can be obtained (see next sections for more details). Among work-first schedulers it seems that the Cilk scheduler is the best except for *FFT* where a *wffl* scheduler gets the best speed-up.

Also, we can observe again the difference in performance depending on the cut-off. In *Alignment* and *SparseLU* there is only a small difference in performance from using one cut-off or the other but in all other applications only schedulers using a particular cut-off are among the top: for *Multisort* and *FFT* cutting by number of tasks works better and for *N Queens* and *Strassen* cutting by the depth level works better. *Alignment* and *SparseLU* are non-recursive applications and this may be the reason why the cutting-off is not so important.

Lesson 3: Beware the Single!. In the *sparseLU* performance results from Figure 2(c) the work-first schedulers do not scale even a bit. The structure of the *sparseLU* application is similar to the one shown in Figure 3. There are a number of untied tasks inside a nest of loops inside a single.

```
1 #pragma omp parallel
2 #pragma omp single
3 #pragma omp task default(shared) untied
4 {
5 for ( k = 0; k < N ; k++ )
6     // bunch of untied tasks
7 }
```

Fig. 3. Structure of a false untied application

All explicit tasks are untied but then the single region forms part of an implicit task. As such, it is always tied. As all tasks (unlike in other applications with a recursive structure) are generated from the single region, but the region cannot threadswitch, a work-first scheduling becomes a serial execution.

This can be easily solved by inserting an extra untied task after the single construct as shown in Figure 4. Now, the generator code is part of an untied task instead of the implicit task so it can be threadswitched.

We have implemented a second version of the *sparseLU* benchmark with this minor modification. We can see, from the results in Figure 5, that the work-first schedulers now achieve a better than the best breadth-first schedule. So, we can see that the effect of this, rather obscure, performance mistake can actually make a good scheduler look bad.

Lesson 4: Deep-First Schedulers Should Be the Default. The *sparseLU* problem is already an indication that the work-first schedulers may have problems when there are no untied tasks. Figure 6 shows how the same schedulers perform when task are untied versus when they are tied.

We can see that in all cases if the tasks are tied performance of the work-first schedulers is severely degraded to the point that no speed-up is obtained.

But, for the breadth-first schedulers the difference is barely noticeable. Moreover, in some cases (*multisort* and *FFT*) the speed-up obtained is better than when tasks are allowed to threadswitch.

Figure 7 shows the average speed-up from all the applications for all combinations of schedulers and cut-offs both for when untied tasks and tied tasks

```
1 #pragma omp parallel
2 #pragma omp single
3 #pragma omp task default(shared) untied
4 {
5 for ( k = 0; k < N ; k++ )
6     // bunch of untied tasks
7 }
```

Fig. 4. Solution to the false untied problem

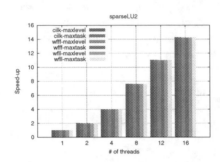

Fig. 5. Speed-ups for sparseLU with an extra untied task

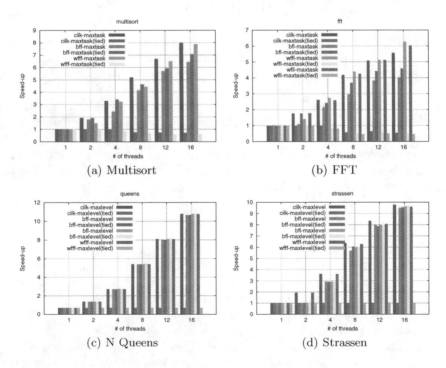

Fig. 6. Untied vs tied tasks

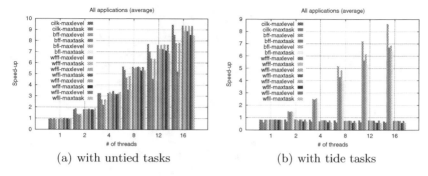

(a) with untied tasks   (b) with tide tasks

Fig. 7. Average speed-ups for all schedulers

are used. These graphs stress our last lesson, when using `untied tasks` work-first schedulers tend to obtain better speed-up but they drop flat when `tied tasks` are used. In that case, breadth-first schedulers perform about the same as they did with `untied task` thus outperforming work-first schedulers.

As `tied` is the OpenMP default, it seems that a wise choice for a compiler (or runtime) is a breadth-first scheduler unless it can safely be guaranteed that all tasks will be `untied`.

5   Conclusions and Future Work

In this work, we have explored several scheduling strategies of OpenMP tasks. We found that while work-first schedulers, in general, obtain better performance that breadth-first schedulers they are not appropriate to be used as the default for OpenMP programs. This is because `tied` and implicit tasks (which may be difficult to spot for novice users) severely restrict the performance that can be obtained. And in those circumstances, which are the default for OpenMP , breadth-first schedulers outperform work-first schedulers.

We have found that while it is a good idea to have a cut-off mechanism, it is not clear which one to use as it may affect negatively the performance of the application and more research is needed in that direction.

As future work, it would be interesting to explore a hybrid cut-off strategy (that takes into account the maximum number of tasks and the depth level) as well as some other more complex cut-off strategies that try to estimate the granularity of the work of a task. Also, it would be interesting to develop a scheduler that detects at runtime the structure of the application (whether it is recursive, whether it uses `tied tasks` or not, . . .) and it chooses one scheduler or the other appropriately.

Acknowledgments

This work has been supported by the Ministry of Education of Spain under contract TIN2007-60625, and the European Commission in the context of the SARC project #27648 and the HiPEAC Network of Excellence IST-004408.

References

1. Acar, U.A., Blelloch, G.E., Blumofe, R.D.: The data locality of work stealing. In: SPAA 2000: Proceedings of the twelfth annual ACM symposium on Parallel algorithms and architectures, pp. 1–12. ACM, New York (2000)
2. Ayguadé, E., Copty, N., Duran, A., Hoeflinger, J., Lin, Y., Massaioli, F., Su, E., Unnikrishnan, P., Zhang, G.: A proposal for task parallelism in OpenMP. In: Proceedings of the 3rd International Workshop on OpenMP, Beijing, China (June 2007)
3. Ayguadé, E., Duran, A., Hoeflinger, J., Massaioli, F., Teruel, X.: An Experimental Evaluation of the New OpenMP Tasking Model. In: Proceedings of the 20th International Workshop on Languages and Compilers for Parallel Computing (October 2007)
4. Balart, J., Duran, A., Gonzàlez, M., Martorell, X., Ayguadé, E., Labarta, J.: Nanos Mercurium: a Research Compiler for OpenMP. In: Proceedings of the European Workshop on OpenMP 2004 (October 2004)
5. Cooley, J.W., Tukey, J.W.: An algorithm for the machine calculation of complex fourier series. Mathematics of Computation 19, 297–301 (1965)
6. Fischer, P.C., Probert, R.L.: Efficient procedures for using matrix algorithms. In: Proceedings of the 2nd Colloquium on Automata, Languages and Programming, London, UK, pp. 413–427. Springer, Heidelberg (1974)
7. Frigo, M., Leiserson, C.E., Randall, K.H.: The implementation of the Cilk-5 multi-threaded language. In: PLDI 1998: Proceedings of the ACM SIGPLAN 1998 conference on Programming language design and implementation, pp. 212–223. ACM Press, New York (1998)
8. Huelsbergen, L., Larus, J.R., Aiken, A.: Using the run-time sizes of data structures to guide parallel-thread creation. In: LFP 1994: Proceedings of the 1994 ACM conference on LISP and functional programming, pp. 79–90. ACM, New York (1994)
9. Korch, M., Rauber, T.: A comparison of task pools for dynamic load balancing of irregular algorithms: Research articles. Concurr. Comput. Pract. Exper. 16(1), 1–47 (2004)
10. Loidl, H.-W., Hammond, K.: On the Granularity of Divide-and-Conquer Parallelism. In: Glasgow Workshop on Functional Programming, Ullapool, Scotland, July 8–10, 1995, Springer, Heidelberg (1995)
11. Mohr, J. E., Kranz, D.A., Halstead, R.H.: Lazy task creation: a technique for increasing the granularity of parallel programs. In: LFP 1990: Proceedings of the 1990 ACM conference on LISP and functional programming, pp. 185–197. ACM, New York (1990)
12. Narlikar, G.J.: Scheduling threads for low space requirement and good locality. In: SPAA 1999: Proceedings of the eleventh annual ACM symposium on Parallel algorithms and architectures, pp. 83–95. ACM, New York (1999)
13. OpenMP Architecture Review Board. OpenMP Application Program Interface, Version 3.0 (Draft) (October 2007)
14. Shah, S., Haab, G., Petersen, P., Throop, J.: Flexible control structures for parallellism in OpenMP. In: 1st European Workshop on OpenMP (September 1999)
15. Teruel, X., Martorell, X., Duran, A., Ferrer, R., Ayguadé, E.: Support for OpenMP tasks in Nanos v4. In: CAS Conference 2007 (October 2007)

Extending the OpenMP Tasking Model to Allow Dependent Tasks

Alejandro Duran, Josep M. Perez, Eduard Ayguadé,
Rosa M. Badia, and Jesus Labarta

Barcelona Supercomputing Center (BSC) - Technical University of Catalunya (UPC)

Abstract. Tasking in OpenMP 3.0 has been conceived to handle the dynamic generation of unstructured parallelism. New directives have been added allowing the user to identify units of independent work (tasks) and to define points to wait for the completion of tasks (task barriers). In this paper we propose an extension to allow the runtime detection of dependencies between generated tasks, broading the range of applications that can benefit from tasking or improving the performance when load balancing or locality are critical issues for performance. Furthermore the paper describes our proof-of-concept implementation (SMP Superscalar) and shows preliminary performance results on an SGI Altix 4700.

1   Introduction

OpenMP grew out of the need to standardize the directive languages of several vendors in the 1990s. It was structured around parallel loops and was meant to handle dense numerical applications. The simplicity of its original interface, the use of a shared memory model, and the fact that the parallelism of a program is expressed in directives that are loosely-coupled to the code, all have helped OpenMP become well-accepted today.

The latest specification released includes tasking, which has been conceived to handle the dynamic generation of unstructured parallelism. This allows programmers to parallelize program structures like `while` loops and recursive functions more easily and efficiently. When a thread in a parallel team encounters a task directive, the data environment is captured. That environment, together with the code represented by the structured block, constitutes the generated task. The data-sharing attribute clauses private, firstprivate, and shared determine whether variables are private to the data environment, copied to the data environment and made private, or shared with the thread generating the task, respectively. The task may be executed immediately or may be queued for execution. All tasks created by a team in a parallel region are completed at the next barrier. It is also possible to wait for all tasks generated by a given task (whether implicit or explicit) using the taskwait directive.

The Intel *work-queueing* model [1] was an early attempt to add dynamic task generation to OpenMP. This proprietary extension to OpenMP allows hierarchical generation of tasks by nesting `taskq` constructs. Synchronization of

R. Eigenmann and B.R. de Supinski (Eds.): IWOMP 2008, LNCS 5004, pp. 111–122, 2008.

descendant tasks is controlled by means of implicit barriers at the end of `taskq` constructs. Tasks have to be defined in the lexical extent of a `taskq` construct.

The Nanos group at UPC proposed *dynamic sections* as an extension to the standard `sections` construct to allow dynamic generation of tasks [2]. Direct nesting of `section` blocks is allowed, but hierarchical synchronization of tasks can only be attained by nesting parallel regions. The Nanos group also proposed the `pred` and `succ` constructs to specify precedence relations among statically named `sections` in OpenMP [3]. [4] also proposed an extension to define a name for `section` and to specify that a `section dependson` another named `section`.

2   Motivation

Task parallelism in OpenMP 3.0 [5] gives programmers a way to express patterns of concurrency that do not match the worksharing constructs defined in the current OpenMP 2.5 specification. The extension in 3.0 addresses common operations like complex, possibly recursive, data structure traversal, and situations which could easily cause load imbalance. However tasking, as currently propose in 3.0, may still be too rigid too express all parallelism available in some applications, specially when the scalability to a high number of cores is the target.

```
1 void fwd(float *diag, float *col);
2 void bmod(float *row, float *col, float *inner);
3 void bdiv(float *diag, float *row);
4 void lu0(float *diag);
5
6 int sparseLU() {
7     int ii, jj, kk;
8
9     for (kk=0; kk<NB; kk++) {
10        lu0(A[kk][kk]);
11        /* fwd phase */
12        for (jj=kk+1; jj<NB; jj++)
13            if (A[kk][jj] != NULL)
14                fwd(A[kk][kk], A[kk][jj]);
15        /* bdiv and bmod phases */
16        for (ii=kk+1; ii<NB; ii++)
17            if (A[ii][kk] != NULL) {
18                bdiv (A[kk][kk], A[ii][kk]);
19                for (jj=kk+1; jj<NB; jj++)
20                    if (A[kk][jj] != NULL)
21                    {
22                        if (A[ii][jj]==NULL) A[ii][jj]=allocate_clean_block();
23                        bmod(A[ii][kk], A[kk][jj], A[ii][jj]);
24                    }
25            }
26    }
27 }
```

Fig. 1. Main code of the sequential SparseLU kernel

To motivate the proposal we use one of the examples that was used to test the appropriateness and performance of the tasking proposal in OpenMP 3.0: the sparseLU kernel shown in Figure 1. This kernel computes an LU matrix

```
1 void fwd(float *diag, float *col);
2 void bmod(float *row, float *col, float *inner);
3 void bdiv(float *diag, float *row);
4 void lu0(float *diag);
5
6 int sparseLU() {
7    int ii, jj, kk;
8
9    for (kk=0; kk<NB; kk++) {
10       lu0(A[kk][kk]);
11 #pragma omp parallel
12 {
13       /* fwd phase */
14 #pragma omp for schedule(dynamic, 1) nowait
15       for (jj=kk+1; jj<NB; jj++)
16          if (A[kk][jj] != NULL)
17             fwd(A[kk][kk], A[kk][jj]);
18
19       /* bdiv phase */
20 #pragma omp for schedule(dynamic, 1)
21       for (ii=kk+1; ii<NB; ii++)
22          if (A[ii][kk] != NULL)
23             bdiv (A[kk][kk], A[ii][kk]);
24
25       /* bmod phase */
26 #pragma omp for schedule(dynamic, 1) private(jj)
27       for (ii=kk+1; ii<NB; ii++)
28          if (A[ii][kk] != NULL)
29             for (jj=kk+1; jj<NB; jj++)
30                if (A[kk][jj] != NULL)
31                {
32                   if (A[ii][jj]==NULL) A[ii][jj]=allocate_clean_block();
33                   bmod(A[ii][kk], A[kk][jj], A[ii][jj]);
34                }
35    }
36 }
37 }
```

Fig. 2. Main code of the OpenMP 2.5 SparseLU kernel

factorization. The matrix is organized in blocks that may not be allocated. In this kernel, once lu0 is computed (line 10), all instances of fwd and bdiv can be executed in parallel (lines 14 and 18, respectively). Each pair of instances fwd and bdiv allow the execution of an instance of bmod (line 23). Across consecutive iterations of the kk loop there are dependences between each instance of bmod and instances of lu0, fwd, bdiv and bmod in the next iteration.

With these data dependences in mind, the programmer could use the current worksharing directives in 2.5 to partially exploit the parallelism available in the kernel, for example using for to distribute the work in the loops on lines 15, 21 and 27 or 29 in Figure 2. Due to the sparseness of the matrix, a lot of imbalance exists, forcing the programmer to use dynamic scheduling of the iterations to have good load balance. For the bmod phase we have two options: parallelize the outer (line 27) or the inner loop (line 29). If the outer loop is parallelized, the overhead is lower but the imbalance is greater. On the other hand, if the inner loop is parallelized the iterations are smaller which allows a dynamic schedule to have better balance but the overhead of the worksharing is much higher.

Notice that it has been necessary to apply loop distribution to isolate the loop that executes the multiple instances of function bdiv. The nowait clause in the loop in line 14 allows the exploitation of the parallelism that exist among

```
 1 void fwd(float *diag, float *col);
 2 void bmod(float *row, float *col, float *inner);
 3 void bdiv(float *diag, float *row);
 4 void lu0(float *diag);
 5
 6 int sparseLU() {
 7    int ii, jj, kk;
 8 #pragma omp parallel
 9    for (kk=0; kk<NB; kk++) {
10 #pragma omp single
11       lu0(A[kk][kk]);
12       /* fwd phase */
13 #pragma omp for nowait
14       for (jj=kk+1; jj<NB; jj++)
15          if (A[kk][jj] != NULL)
16 #pragma omp task firstprivate(kk, jj)
17             fwd(A[kk][kk], A[kk][jj]);
18       /* bdiv phase */
19 #pragma omp for
20       for (ii=kk+1; ii<NB; ii++)
21          if (A[ii][kk] != NULL)
22 #pragma omp task firstprivate(kk, ii)
23             bdiv (A[kk][kk], A[ii][kk]);
24
25       /* bmod phase */
26 #pragma omp for private(jj)
27       for (ii=kk+1; ii<NB; ii++)
28          if (A[ii][kk] != NULL)
29             for (jj=kk+1; jj<NB; jj++)
30                if (A[kk][jj] != NULL)
31 #pragma omp task firstprivate(kk, jj, ii)
32                {
33                   if (A[ii][jj]==NULL) A[ii][jj]=allocate_clean_block();
34                   bmod(A[ii][kk], A[kk][jj], A[ii][jj]);
35                }
36    }
37 }
```

Fig. 3. Main code of SparseLU with OpenMP 3.0 tasks

the instances of functions `fwd` and `bdiv`. The implicit barrier at the end of worksharing in line 20 forces the dependences of `fwd` and `bdiv` with `bmod`.

Using the task proposed in 3.0, the code restructuring is quite similar, as shown in Figure 3; however tasks allow to only create work for non-empty matrix blocks. We also create smaller units of work in the bmod phase with an overhead similar to the outer loop parallelization. This reduces the load imbalance problems. The `nowait` clause in line 13 allows the parallel execution of `fwd` and `bdiv` instances. The implicit barriers at the end of loops in lines 19 and 16 force the dependences between pairs of `fwd`/`bdiv` with `bmod` inside a single `kk` iteration and viceversa across consecutive iterations of loop `kk`.

Figure 4 shows an execution trace obtained from an instrumented run of the kernel and visualized with Paraver [6]. The window represents time in horizontal axis and per-thread activity in the vertical axis (in this case, each color identifies the function that is being executed). The visualization corresponds to the end of a `kk` iteration and the beginning of the next `kk+1` iteration. Yellow lines represent thread creation and thread execution points (in the window only for `fwd` and `bdiv`).

As we pointed at the beginning of this section, there exists more parallelism in this kernel that can not be exploited with the current task definitions: parallelism that exists between tasks created in lines 17 (fwd) and 23 (bdiv) and tasks

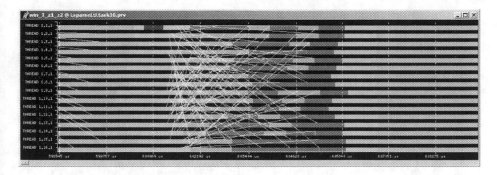

Fig. 4. Paraver window with a portion of SparseLU execution: lu0 (blue), fwd (green), bdiv (red) and bmod (orange) functions

created in line 34 (bmod) inside a single iteration . Also it would be interesting to express the parallelism that exists across consecutive iterations of the kk loop.

3   Proposed Extension

In this section we describe the extensions we propose to the OpenMP tasking model. We first describe them as part of the *StarSs* framework, a new programming paradigm for task-based programming that targets homogeneous symmetric multiprocessors (*SMPSs*) and the Cell/B.E. architecture [7] (*CellSs* [8]).

3.1   StarSs Pragmas and Execution Model

With *StarSs* the programmer identifies the functions that will be executed as tasks, using a pragma annotation right before the function definition. In addition the programmer specifies the directionality of each of the function parameters: input, output or input/output.

```
#pragma smpss task [clause[[,]clause] ...]
          {function-header|function-declaration}
```

where clauses can be:

- input(argument-list)
- output(argument-list)
- inout(argument-list)

Each element in `argument-list` is a block of contiguous memory locations whose number of elements is specified either in the function header or in the construct.

The following optional pragmas indicate a scope of the program where StarSs is used:

```
#pragma smpss start
#pragma smpss finish
```

When the `start` pragma is reached, the runtime initializes a worker thread in each processing element, who will wait for tasks to execute. Only a single thread (main thread) continues with the execution of the program, dynamically creating the tasks that are stored in a task graph. Both the main thread and the worker threads get tasks from the task graph once dependences are honored and execute the function associated. The `finish` pragma finishes all idle threads once the task graph is totally executed. Functions annotated with task have to be called between these two pragmas. If they are not present in the user code, the compiler will automatically insert the start pragma at the beginning of the application and the finish pragma at the end.

Figure 5 shows the SparseLU kernel programmed with the SMPSs extensions. The programmer identifies four tasks that correspond to the execution of functions `lu0`, `fwd`, `bdiv` and `bmod`. For example, for function `bmod` the programmer is specifying that the first and second arguments (`row` and `col`) are input parameters (they are only read during the execution of the function) and that the third argument (`inner`) is inout since it is read and written during the execution of the function. Notice that the annotations are placed on the original

```
1 #pragma smpss task input(diag[B][B]) inout(col[B][B])
2 void fwd(float *diag, float *col);
3
4 #pragma smpss task input(row[B][B],col[B][B]) inout(inner[B][B])
5 void bmod(float *row, float *col, float *inner);
6
7 #pragma smpss task input(diag[B][B]) inout(row[B][B])
8 void bdiv(float *diag, float *row);
9
10 #pragma smpss task inout(diag[B][B])
11 void lu0(float *diag);
12
13 int sparseLU() {
14     int ii, jj, kk;
15
16 #pragma smpss start
17     for (kk=0; kk<NB; kk++) {
18         lu0(A[kk][kk]);
19         /* fwd phase */
20         for (jj=kk+1; jj<NB; jj++)
21             if (A[kk][jj] != NULL)
22                 fwd(A[kk][kk], A[kk][jj]);
23         /* bdiv and bmod phases */
24         for (ii=kk+1; ii<NB; ii++)
25             if (A[ii][kk] != NULL) {
26                 bdiv (A[kk][kk], A[ii][kk]);
27                 for (jj=kk+1; jj<NB; jj++)
28                     if (A[kk][jj] != NULL)
29                     {
30                         if (A[ii][jj]==NULL) A[ii][jj]=allocate_clean_block();
31                         bmod(A[ii][kk], A[kk][jj], A[ii][jj]);
32                     }
33             }
34     }
35 #pragma smpss finish
36 }
```

Fig. 5. Main code of SparseLU with StarSs tasks

sequential version, with no transformations applied to allow the specification of the inherent parallelism available.

When a call to a function annotated with the `task` construct is found, the main thread creates a task for the associated function and adds information about data dependencies in a task graph. For each task, the runtime dynamically computes data dependencies by analyzing the direction (input, output or both), length and address of each parameter against those of previous tasks in sequential order. *True* data dependences (read-after-write) are honored by the runtime system by deferring the execution of the task until all `input` and `inout` arguments have been computed. The execution of the task can be done by any thread in the current parallel team. Once a task finishes its execution, the runtime updates the task graph to signal the modification of all `output` and `inout` arguments.

The runtime system automatically removes *false* dependencies (write-after-read and write-after-write) using memory renaming, a technique borrowed from the idea of register renaming in current out-of-order superscalar processors. For each variable that needs to be renamed, the runtime allocates temporary memory space for it. That is, if a task writes to an array, renaming can replace that array by a temporary one and redirect all following reads of that definition to the temporary array.

While the underlying runtime is capable of handling all inter-task related data dependencies, it cannot handle dependencies with the code executed by the master thread. To handle this, StarSs includes a data barrier:

```
#pragma smpss wait on (address-list)
```

At the `wait on` pragma, the master thread waits for all memory locations in the `address-list` to be updated. Once this happens, the main thread continues with the execution of the code.

3.2   StarSs and OpenMP

The StarSs pragmas and execution model fit well with the tasking definition in OpenMP 3.0

```
#pragma omp task [clause[[,]clause] ...]
        structured-block
```

In addition to the clauses supported in OpenMP 3.0:

- untied
- shared (variable-list)
- firstprivate (variable-list)
- private (variable-list)

our proposal is to include:

- input(variable-list)
- output(variable-list)
- inout(variable-list)

```
1 int sparseLU() {
2    int ii , jj , kk;
3
4 for (kk=0; kk<NB; kk++) {
5 #pragma omp task inout(A[kk][kk])
6    lu0(A[kk][kk]);
7    for (jj=kk+1; jj<NB; jj++)?
8       if (A[kk][jj] != NULL)?
9 #pragma omp task input(A[kk][kk]) inout(A[kk][jj])
10         fwd(A[kk][kk], A[kk][jj]);
11
12    for (ii=kk+1; ii<NB; ii++) {
13       if (A[ii][kk] != NULL)?
14 #pragma omp task input(A[kk][kk]) inout(A[ii][kk])
15         bdiv (A[kk][kk], A[ii][kk]);
16         for (jj=kk+1; jj<NB; jj++)?
17            if (A[kk][jj] != NULL) {
18               if (A[ii][jj]==NULL) A[ii][jj]=allocate_clean_block();
19 #pragma omp task input(A[ii][kk], A[kk][jj]) inout(A[kk][kk])
20               bmod(A[ii][kk], A[kk][jj], A[ii][jj]);
21            }
22       }
23 }
```

Fig. 6. Main code of SparseLU with the proposed dependent tasks, version 1

We also propose to include the

`#pragma omp wait on (address-list)`

in order to provide a more flexible version of `taskwait`.

The first difference wit StarSs is that our proposed clauses apply to an OpenMP task, which is a structured block of code and not a function declaration or definition. The main implication of this is that the `variable-list` does not indicate formal function arguments but variables used in the scope of the structured block of code. Figure 6 shows the SparseLU example with the proposed extension in OpenMP.

The second difference is that StarSs forces dependent tasks to be generated in sequential order (or at least in an order that guarantees that the source is generated before the target of the dependence). In addition, only the *main thread* can generate tasks for the *worker* threads. In OpenMP is it possible to have multiple task generators (by having `task` inside a worksharing or by nesting `task`). This needs to be considered in the implementation of the extensions in the prototype OpenMP implementation, but in any case, it is the programmer responsibility to ensure the appropriate order of task generation.

Clauses `Input`, `output` and `inout` provide additional information to the `shared` data clause. This information is used by the runtime to dynamically build and update the task graph and schedule tasks for execution as soon as all their input variables are generated. A variable in a `shared` data clause, but not in a `input`, `output` or `inout` clause, indicates that the variable is accessed inside the task but it is not affected by any data dependence in the current scope of execution (or is protected by another one). `Firstprivate` variables could also be affected with an `input` clause, meaning that the per-task private copy of the variable should be initialized with the value generated by another task (in its `output` clause) instead of the value at creation time.

```
1 int sparseLU() {
2     int ii , jj , kk;
3     int lu0done , fwddone[NB] , bdivdone[NB] , bmoddone[NB][NB];
4
5 for (kk=0; kk<NB; kk++) {
6 #pragma omp task input(bmoddone[kk][kk]) output(lu0done)
7     lu0(A[kk][kk]);
8     for (jj=kk+1; jj<NB; jj++)?
9         if (A[kk][jj] != NULL)?
10 #pragma omp task input(lu0done,bmoddone[kk][kk]) output(fwddone[jj])
11         fwd(A[kk][kk], A[kk][jj]);
12
13     for (ii=kk+1; ii<NB; ii++) {
14         if (A[ii][kk] != NULL)?
15 #pragma omp task input(lu0done,bmoddone[kk][kk]) output(bdivdone[ii])
16         bdiv (A[kk][kk], A[ii][kk]);
17             for (jj=kk+1; jj<NB; jj++)?
18             if (A[kk][jj] != NULL) {
19                 if (A[ii][jj]==NULL) A[ii][jj]=allocate_clean_block();
20 #pragma omp task input(bdivdone[ii],fwddone[jj]) inout(bmoddone[kk][kk])
21                 bmod(A[ii][kk], A[kk][jj], A[ii][jj]);
22             }
23         }
24 }
```

Fig. 7. Main code of SparseLU with the proposed dependent tasks, version 2

Previous proposals based on providing a name to each **section** or **task** [3,4] can also be implemented using the proposed extensions in this paper, as shown in Figure 7. In this case, a dependence is encapsulated in a variable that should be declared by the programmer and used in an **output** clause (in the source task) and and in a **input** clause (in the target task). This synchronization variable can be subject of reuse and therefore, false dependences; the automatic renaming mechanism in the runtime avoids these false dependences and avoids its scalar (or vector) expansion.

4   Additional Runtime Features

The prototype task implementation for OpenMP 3.0 enqueues new created tasks in a team pool of tasks. Any thread of the team can access this pool a execute the tasks from there. Threads have also a local pool in which they place those tasks that have been suspended by them if those tasks are *tied tasks*. Other threads are not allowed to steal tasks from this pool. But the OpenMP specification allows for other forms of scheduling (with certain restrictions related to tied/untied tasks). For example, it would be possible to implement a work-first scheduler (like Cilk [9] does) where tasks are executed as soon as they are created and the parent task is suspended and stored in a per task pool of tasks. Dependence restrictions would need to be considered in this case. To avoid starvation (because all tasks go to the local pools) work-stealing is allowed.

In the implementation of SMPSs each thread has a local pool of ready tasks. The main thread is responsible of running the main program by going through the non task user code, analyzing the data dependencies and adding the tasks to the task graph. New tasks that have no input dependencies are added to the main thread task pool; any worker thread can steal from the pool of the main thread. When the main thread stops task generation (because the task pool is full or he is waiting for tasks to finish) it also execute tasks from its own pool.

Worker threads look for ready tasks first in their own pool, then on the main thread pool and then on the other thread pools. When a thread finishes running a task, it puts all the task successors that have become ready into its task pool. While worker threads consume tasks from their pool in LIFO order, they steal them from other threads in FIFO order. That is, they consume the graph in a depth first order as long as they can can get ready tasks, and then steal tasks from other threads in a breadth first order when their task pools become empty.

The idea behind this design is that each thread will be executing tasks in a different region of the graph and have little interference with other threads as long as there are ready tasks in that region or there are unexplored zones in the graph. Otherwise they will steal work from other threads in a way that tries to minimize the effect on the cache locality of that thread.

5   Preliminary Evaluation

In order to test the proposal in terms of expressiveness and performance, we have developed the StarSs runtime for SMP (named SMPSs) and used the Mercurium compiler (source-to-source restructuring tool) [2]. For comparison purposes we also use the reference implementation [10] of the tasking proposal in OpenMP 3.0 based on the Nanos runtime and the same source-to-source restructuring tool. and the workqueueing implementation available in the Intel compiler.

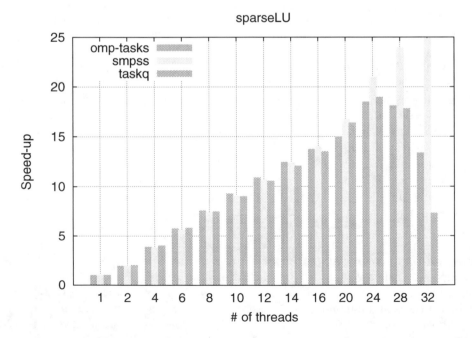

Fig. 8. Speed-up of *taskq*, *task* and *smpss* for SparseLU

We evaluate how the proposed extension improves the scalability of the SparseLU benchmark that has been used to motivate the proposal. All the executions have been done on an SGI Altix 4700 using up to 32 processors in a cpuset (to avoid interference with other running applications).

Figure 8 shows the speed-up with respect to the sequential execution time. Notice that up to 16 threads the three versions (*taskq*, *task* and *smpss*) behave similarly. When more threads are used, load unbalancing starts to be more noticeable and the overheads of tasking are not compensated with the parallel execution. Task barriers between *fwd/bdiv* and *bmod* phases (inside iteration kk) and between *bmod* and *fwd/bdiv* phases (in consecutive iterations of kk) introduce this load unbalance and overheads. However, *smpss* is able to overcome these two limitations by overlapping tasks in these computational phases inside and across iterations of the kk loop.

The implementation of SMPSs has overheads. Table 1 shows a breakdown of the execution time of the *SMPSs* version of SparseLU. The table shows the percentage of time that each thread is in each phase (*worker threads'* information has been summarized due to space limitations). For this example, the *main thread* invests around the 30% of its time in the maintenance of the task graph, and around 65 % of its time is left for execution of tasks. The worker threads also suffer of some overheads (around 5%), not only due to the maintenance of the task graph but also to the time the threads are waiting for tasks ready to be executed and the time invested in getting the tasks description. Depending on the application and on the number of threads, these overheads will have more or less impact in the performance, but we are looking for more efficient implementations of the task graph to reduce them.

Table 1. Breakdown of SMPSs overheads for the SparseLU with 16 threads

Thread phase	Main Thread	Max Worker th.	Min Worker th.	Avg. Worker th.
User code	5.12 %			
Initialisation	0.13 %			
Adding task	10.51 %			
Remove tasks	19.67 %	2.41 %	0.86 %	1.46 %
Waiting for tasks	0.46 %	1.95 %	1.04 %	1.47 %
Getting task descr.	0.36 %	1.28 %	0.56 %	1.10 %
Tasks' execution	63.76 %	97.43 %	94.97 %	95.97 %

6   Conclusions

This paper proposed an extension to the OpenMP 3.0 tasking model: data dependent tasks. Data dependencies among tasks are indirectly expressed by specifying the input and output direction of the arguments used in a task. This is a key difference with respect to previous proposals that were based on the specification of named tasks and **dependson** relationships.

The paper uses one of the application kernels used to demonstrate the expressiveness of tasking in OpenMP 3.0: SparseLU. We motivate the proposal with this kernel and show how its scalability improves with a prototype implementation of the proposal (SMP Superscalar – SMPSs).

The possibility of expressing input and output direction for the data used by the task provides extra benefits for other multicore architectures, such as for example the Cell/B.E. processor [7] (Cell Superscalar [8]). In this case, the information provided by the programmer allows the runtime system to transparently inject data movement (DMA transfers) between SPEs or between SPEs and main memory.

Acknowledgments

The Programming Models group at BSC-UPC is supported by the IBM MareIncognito project, the European Commission in the context of the SARC project (contract no. 27648) and the HiPEAC Network of Excellence (contract no. IST-004408), and the Spanish Ministry of Education (contracts no. TIN2004-07739-C02-01 and TIN2007-60625).

References

1. Shah, S., Haab, G., Petersen, P., Throop, J.: Flexible control structures for parallelism in OpenMP. In: 1st European Workshop on OpenMP (September 1999)
2. Balart, J., Duran, A., Gonzàlez, M., Martorell, X., Ayguadé, E., Labarta, J.: Nanos mercurium: a research compiler for openmp. In: Proceedings of the European Workshop on OpenMP 2004 (October 2004)
3. Gonzàlez, M., Ayguadé, E., Martorell, X., Labarta, J.: Exploiting pipelined executions in OpenMP. In: 32nd Annual International Conference on Parallel Processing (ICPP 2003) (October 2003)
4. Sinnen, O., Pe, J., Kozlov, A.: Support for Fine Grained Dependent Tasks in OpenMP. In: 3rd International Workshop on OpenMP (IWOMP 2007) (2007)
5. Ayguadé, E., Copty, N., Duran, A., Hoeflinger, J., Lin, Y., Massaioli, F., Unnikrishnan, P., Zhang, G.: A Proposal for Task Parallelism in OpenMP. In: 3rd International Workshop on OpenMP (IWOMP 2007) (2007)
6. Labarta, J., Girona, S., Pillet, V., Cortes, T., Gregoris, L.: Dip: A parallel program development environment. In: 2nd International Euro-Par Conference on Parallel Processing (1996)
7. Pham, D., Asano, S., Bolliger, M., Day, M.N., Hofstee, H.P., Johns, C., Kahle, J., et al.: The Design and Implementation of a First-Generation Cell Processor. In: IEEE International Solid-State Circuits Conference (ISSCC 2005) (2005)
8. Bellens, P., Perez, J.M., Badia, R.M., Labarta, J.: CellSs: a programming model for the Cell BE architecture. In: proceedings of the ACM/IEEE SC 2006 Conference (November 2006)
9. Frigo, M., Leiserson, C.E., Randall, K.H.: The implementation of the Cilk-5 multithreaded language. In: PLDI 1998: Proceedings of the ACM SIGPLAN 1998 conference on Programming language design and implementation, New York, pp. 212–223. ACM Press, New York (1998)
10. Ayguadé, E., Duran, A., Hoeflinger, J., Massaioli, F., Teruel, X.: An experimental evaluation of the new openmp tasking model. In: Proceedings of the 20th International Workshop on Languages and Compilers for Parallel Computing (October 2007)

OpenMP Extensions for Generic Libraries

Prabhanjan Kambadur, Douglas Gregor, and Andrew Lumsdaine

Open Systems Laboratory, Indiana University
{pkambadu,dgregor,lums}@osl.iu.edu

Abstract. This paper proposes extensions to the OpenMP standard to provide first-class support for parallelizing generic libraries such as the C++ Standard Library (SL). Generic libraries are especially known for their efficiency, reusability and composibility. As such, with the advent of ubiquitous parallelism, generic libraries offer an excellent avenue for parallelizing the existing applications that use these libraries without requiring the applications to be rewritten. OpenMP, which would be ideal for executing such parallelizations, does not support many of the modern C++ idioms such as *iterators* and *function objects* that are used extensively in generic libraries. Accordingly, we propose extensions to OpenMP to better support modern C++ idioms to aid in the parallelization of generic libraries and applications built with those libraries.

1 Introduction

For many years, the performance of mainstream software applications was able to improve at a fairly rapid pace because the performance of CPU cores increased more or less in accordance with Moore's law. Although Moore's Law is still in effect (VLSI feature sizes are still decreasing), recent increases in chip density do not come with corresponding increases in clock speeds or single processor performance. Rather, more processing cores are being put onto single chips, with the result that now even mainstream applications must look to parallel programming in order to obtain continued increases in performance [1,15].

Parallelizing mainstream applications obviously requires that parallelism be somehow expressible in mainstream programming languages. Approaches to providing such support include language extensions, libraries, compiler directives, and automatic parallelizing compilers. There are a number of issues that come into play in determining how effective any of these approaches will be. The jury is still out as to which approaches will be most effective—the programming community needs to gain experience with a wide variety of possible approaches to determine the best one. Indeed, there may not be a best approach and some combination of approaches will need to be available to programmers indefinitely.

To become accepted by mainstream programmers, parallelization technologies must account for established programming practices. For example, beyond issues of programming languages, modern software development is based on key programming paradigms (e.g., the object-oriented paradigm or the generic programming paradigm) and the extensive use of libraries. In many cases, it will

R. Eigenmann and B.R. de Supinski (Eds.): IWOMP 2008, LNCS 5004, pp. 123–133, 2008.

be established sequential codes that must be parallelized. New approaches to parallelism must therefore be unobtrusive and support incremental adoption.

OpenMP is a well-understood and mature parallelization technology [3]. Although OpenMP currently provides some support for the C++ language, the full range of paradigms and libraries that are common to C++ are not well-supported. In particular, the generic programming paradigm has recently emerged as an important development practice for C++. Generic programming in C++ simultaneously stresses modularity and performance and has garnered widespread support in the C++ community and standardization bodies. Currently there is little built-in support for parallelism in the standardized C++ language and this situation is unlikely to change in the foreseeable future. Rather, support for parallelism in C++ programs will need to come from libraries (e.g., Boost.MPI [8] or the Threading Building Blocks [11]) or from compiler directives such as those provided by OpenMP.

Accordingly, this paper presents a proposal for extensions to OpenMP to support modern generic programming practice in C++. This proposal results from a decade of experience with generic programming (and even longer with parallel programming). Based on the analysis of the generic programming process and of several important C++ libraries (reported in [7]), we identified the areas where new functionality is required from OpenMP in order for it to properly support generic programming. In particular, our proposal includes extensions to the for and reduction clauses and introduces a new clause, requires. Our proposed extensions follows the guiding principles of OpenMP: they are simple to use, simple to specify, and consistent with rest of OpenMP.

Direct support of generic programming in the proposed fashion will greatly facilitate parallelization of modern C++ libraries and applications. In fact, parallelized generic algorithms retain their sequential semantics. Programmers will be able to realize immediate (and seamless) parallelization of their applications simply by using parallelized libraries. Finally, and importantly with respect to evolving the practice of parallel programming, our proposed approach provides a disciplined conceptual framework in which to reason about parallelism.

2   Motivation

We motivate our proposal with an extended example based on the generic accumulate algorithm and use it to illustrate the issues that must be addressed in extending OpenMP for use with generic programming in C++. The accumulate algorithm is a generalization of summation over all the elements in a collection and is a stereotypical example of many of the algorithms in the C++ Standard Library (SL). The SL version of accumulate is as follows:

```
template<typename InputIterator, typename T, typename BinaryFunction>
T accumulate (InputIterator first, InputIterator last, T init, BinaryFunction f) {
      for ( ; first != last; ++first) init = f(init, *first);
}
```

Note that technically accumulate is a *function template*, not a function. That is, rather then being expressed in terms of concrete types, it is paramaterized by three template parameters: InputIterator, T and BinaryFunction. When the accumulate algorithm is invoked, the template parameters are bound to actual types and the resulting instantiated algorithm is compiled. Thus, the very same accumulate algorithm can be used with arrays, linked lists, or many other types (but not *any* other type, as we will see below):

```
double x[10];
double a = accumulate(x, x+10, 0.0, plus<double>()); // add elements in a

node* n = cons(1, cons(2, cons(2, null )));
int s = accumulate(n, null, 1, multiplies <int>()); // multiply  elements in n
```

Although originally popularized by the Standard *Template* Library [10], generic programming is about much more than programming with templates. Fundamentally, generic programming is a systematic approach to classifying entities within a problem domain according to their underlying semantics and behaviors. Component interface definitions are based on finding *minimal* interface requirements, thereby providing *maximal* opportunities for composition and reuse. Moreover, because component interfaces are based on requirements rather than on types, components from separate libraries can readily be composed, even if the libraries were developed independently of each other.

The specification of generic libraries thus includes specification of interface requirements (encapsulated as "concepts") as a central feature. The arguments to accumulate must therefore meet the following requirements:

- The type of the parameters first and last must meet the requirements of the InputIterator concept. A type that meets these requirements (said to "model" the concept) can iterate through the elements of a collection. It can be a built-in type, such as double* or an aggregate type such as an iterator over a user-defined container. As can be seen from the body algorithm, the InputIterator concept requires (at least) operator++, operator!=, operator*.
- The type of f must model the BinaryFunction concept, i.e., be a type that has operator() defined for it. Furthermore, operator() accepts two parameters and returns a value that can be converted (and hence assigned) to the type T.
- The final requirement then is that the type bound to T be convertible to the type obtained when InputIterator is dereferenced. Also, the type bound to T must model Assignable, i.e., it must have operator= defined for it, and it must model CopyConstructible, i.e., we must be able to create copies of variables of that type.

The example invocations of accumulate shown above are correct with respect to the above specified requirements. For the first example, double* models InputIterator, std::plus<double> models BinaryFunction, and double can be assigned to return value of std::plus<double>. By expressing accumulate in terms of the InputIterator and BinaryFunction concepts, we have separated the expression of the algorithm from the details of the types upon which the algorithm can operate, a fundamental tenet of generic programming [9].

Now, let us consider parallelizing accumulate using OpenMP. This task might seem straightforward since accumulate is comprised of a for loop that performs a reduction operation. However, the loop is not in the canonical form that is required by OpenMP. To address this (seemingly syntactic) issue, we could rewrite the loop of accumulate as follows (including also the OpenMP pragma):

```
template<typename InputIterator, typename T, typename BinaryFunction>
T accumulate (InputIterator first, InputIterator last, T init, BinaryFunction f) {
        #pragma omp parallel for
        for (int i=0; i < (last−first); ++i) init = f (init, first[i]);
}
```

Although the syntactic form of this loop now lends itself to being parallelized using OpenMP, this form is both incomplete and incorrect. First, note that accumulate is performing a reduction operation and hence the parallel version as given above is incorrect. We must either further modify the algorithm by introducing manual reduction before and after the loop or use OpenMP's reduction clause (about which more will be said). Second, by rewriting the loop and making it parallel, we have introduced additional requirements on the input types:

- The type bound to InputIterator must support operator[] and operator−.
- The operation bound to BinaryFunction must be associative since the operations may be reordered during parallel execution. Also, since OpenMP supports reductions only on certain operations [3], BinaryFunction is required to be an operation that we can map onto an operation supported by OpenMP.
- The type bound to T must posses an *identity* element with respect to the type bound to BinaryFunction so that private variables of T can be properly initialized. Furthermore, since OpenMP reduction clause supports only built-in types, we require that T be a built-in type.

If we create a parallel version of accumulate in this way, we have enabled parallelism but we have broken genericity. The parallel version of accumulate can be invoked only on those types that meet the old requirements *and* that meet the new requirements. It cannot be used on types that only meet the old requirements but not the new—even though, according to the SL specification, that would be a correct invocation of accumulate. The flexibility of the template system in C++ does provide an escape hatch of sorts for this problem, through *algorithm specialization*. Algorithm specialization allows us to take advantage of extra properties that sets of types might have, in order to achieve improved performance (or, in this case, parallelization), by providing overloads of the algorithm based on concept properties. Algorithm specialization is realized with template metaprogramming techniques with the current C++ but is a likely feature for the upcoming C++ standard, C++0x, discussed below.

Thus, to parallelize accumulate, we would need to create a specialization of it with the additional requirements given above. However, we need to write a new specialization for every single reduction operation that is supported by the the OpenMP reduction clause. This measure effectively reduces the *genericity* of the accumulate algorithm since each specialization is *bound* to a particular type of

BinaryFunction. It is also a poor use of specialization: since the accumulate algorithm itself remains the same (only the pragma and requirements differ), this approach requires significant redundant code, which violates basic software development principles. Furthermore, we have had to intrusively change the structure of the accumulate algorithm to conform to OpenMP's canonical form which directly violates OpenMP design principles. Finally, this solution is still incomplete since there is no support for user-defined types or user-defined reduction operations, further limiting its genericity.

Generic Programming with ConceptC++. Currently, generic programming is realized in C++ through various techniques such as polymorphic containers, iterators, function objects and traits classes [4]. Recently, the *concepts* extension to C++ [5] was proposed, which provides complete linguistic support for generic programming, and which is a likely candidate for adoption in the next C++0x standard [6]. One of the main features of concept-enabled C++ ("ConceptC++") is the ability to specify concepts, define algorithms in terms of concepts, and to check whether particular types model specific concepts. Writing generic algorithms and their specializations is greatly simplified as a result. For example, a specialization of the accumulate algorithm for the reduction operation + is given below.

```
template<typename Iterator, typename T, typename Op>
requires RandomAccessIterator<Iterator> && Monoid<Op, T>
         && AddOperation<Op> && BuiltInIntegralType<T>
T accumulate (Iterator first, Iterator last, T init, Op f) {
    #pragma omp parallel for reduction(+:init)
    for (int i=0; i < (last−first); ++i) init = f (init, first[i]);
}
```

Here, the requires clause describes the requirements on the template parameters, e.g., the Iterator type must meet the requirements of the RandomAccessIterator concept. The requires clause both enables type checking of the body of the function template and also ensures that users of the function template have provided types that meet the requirements in the requires clause. The following concepts are contained in the above requires clause:

- RandomAccessIterator<Iterator>: The Iterator type must provide random access to the elements in its collection. Hence, operator[] and operator− are valid on it.
- Monoid<Op, T>: A pair (Op,T) modeling this concept ensures the presence of an identity element of type T with respect to Op. Since Monoid concept also includes all the requirements for (i.e., "refines") the BinaryFunction concept, it guarantees that Op models BinaryFunction as well.
- AddOperation<Op>: This concept refines the BinaryFunction concept and requires Op to be an addition operation.
- BuiltInIntegralType<T>: The type bound to T must model BuiltInIntegralType, meaning it must be a built-in integral type.

Related Work. The first attempt to study the requirements behind parallelizing generic libraries was made by Austern et al. [2] and was based on the IRIS Power C compiler pragmas. The Intel *workqueueing* [12] model enables parallelization of certain iterator-based algorithms by treating each loop statement as a task to be executed. Terboven et al. [16] exposit on some of the shortcomings of OpenMP with respect to modern C++ programming paradigms and suggest certain strategies to overcome these limitations. However, these strategies are aimed at specific problems rather than the fundamentals of generic programming. Finally, Suess et al. [13,14] have reported on shortcomings in implementing sorting algorithms and singleton patterns in C++.

3   Extensions to OpenMP

We propose three primary extensions to OpenMP to provide first-class support for parallelizing generic libraries: extensions to for to support iterators; extensions to reduction to support user-defined types, function objects and overloaded operators; the introduction of the requires clause to support conditional parallelization.

3.1   The for Construct

The OpenMP for loop construct places certain restrictions on the loop form (called the canonical form). This restriction ensures accurate determination of the loop parameters at compile time. However, it also precludes the parallelization of C++ iterator loops despite the fact that some of these iterator loops can be rewritten in the required canonical form [16]. We propose to extend the for loop construct as follows:

Syntax for (*init-expr*; *cond-expr*; *incr-expr*) *statement*
init-expr is required to be in one of the following three forms.

  - ;, the empty expression. In this case, the *lhs* in the *cond-expr* is taken to be the loop iterator.
  - *iter_type iterator = init;*
  - *iterator = init;*
    - *iterator* models the RandomAccessIterator concept.
    - *init* is an expression that returns a value convertible to *iterator*'s type.

cond-expr is of the form *iterator rel-op bound-expr*, where

  - *iterator* is the same variable that was used in *init-expr*. If *init-expr* was an empty statement, then the *iterator* is taken to be the loop variable.
  - *rel-op* is one of <, <=, >, >=, !=. The only addition here is the != operator that is used frequently in generic libaries. Inclusion of this operator does not impede the ability to deduce the iteration space.
  - *bound-expr* is loop-invariant and returns a value that is convertible to the type of *iterator*.

incr-expr is of the form *incr-expr-1[,incr-expr-2,...,incr-expr-n]* where *incr-expr-i* is a currently allowed increment expression. Any iterators named in the *incr-expr* list are required to model the RandomAccessIterator as well.

Use case. The SL copy algorithm copies elements from the range [first, last) to the range [result, result + (last − first)). The return value is (result+(last−first)). Parallelization is possible only when both InputIterator and OutputIterator model the RandomAccessIterator concept.

```
template <typename InputIterator, typename OutputIterator>
OutputIterator copy (InputIterator first, InputIterator last, OutputIterator result) {
#pragma omp parallel for
  for ( ; first!=last; ++first, ++result) *result = *first;
}
```

3.2   The Reduction Clause

We propose to extend the OpenMP reduction clause to support user-defined types, overloaded operators, and function objects in the same way that it currently supports built-ins. Two important requirements for parallel reduction are that the operation be associative and that the operand have an identity element with respect to the operation.

Syntax reduction (*operation*: *operand-1[,operand-2,..,operand-n]*)
operation is either an operator specified by the OpenMP 3.0 standard (built-in or overloaded) or a function object that satisfies the following requirements:

- Model the BinaryFunction or the UnaryFunction concept, and
- Types modeling BinaryFunction must also model the Associative concept.

Note that the operations are not required to be either commutative or re-entrant. However, if the operation does model the Commutative or the ReEntrant concept, additional optimizations may be performed.
operand must satisfy the following:

- Either be a builtin type or an aggregate type (*struct, class*),
- Model the CopyConstructible concept, and
- Model the Assignable concept.

Identity requirement

- Each pair of *(operation, operand-i)* must model the Monoid concept.

Use case With the proposed extensions, the accumulate algorithm described in section 1. can be parallelized as follows:

```
template <typename InputIterator, typename T, typename BinaryFunction>
T accumulate (InputIterator first, InputIterator last, T init, BinaryFunction f) {
#pragma omp parallel for reduction(f:init)
  for ( ; first!=last; ++first) init = f (init, *first);
}
```

To illustrate the ramifications of this version of accumulate, consider its use with the std::string class and the std::plus function object. std::plus is a synonym for the + operator, and models both the BinaryFunction and Associative concepts; in the case of std::string it performs string concatenation (and is therefore not commutative). We establish the relationships between these types and their concepts using the concept_map keyword in ConceptC++. The concept map specifies the empty string as the identity element for the std::plus operation applied to strings. Note that we are using std::vector iterators, which model RandomAccessIterator.

```
concept_map Monoid<plus<string>, string> {string identity() {return string(""); }};

vector<string> string_vec(n);
string init("CONCATENATE");
accumulate (string_vec.begin(), string_vec.end(), init, plus<string>());
```

3.3   The Requires Clause

As discussed in Section 2, parallelization of generic algorithms often imposes additional requirements on its input types, thereby requiring more refined concepts. To meet the principles of generic programming and OpenMP, we need to be able to parallelize a generic algorithm with a single directive and without modifying the generic algorithm. At the same time, since the parallelization will only be correct for more refined concepts, we need a means to express these additional constraints within the scope of the directive. Parallelization will occur conditionally, when the input types model the more refined concepts. Otherwise, the sequential algorithm is instantiated. To enable the expression of conditional parallelization based on concepts, we propose the requires clause.

Syntax requires (*default*)
requires (*concept-name:var-1[,var-2,...,var-n]*)
requires (*cond-expr*)

  – *var-1[,var-2...,var-n]* is the list of function objects, operators or variable declarations that need to model the *concept-name* concept.
  – *cond-expr* is any expression that returns value that is convertible to bool.

Scope. The requires clause may be used with the parallel, for, and sections constructs.

Behavior. When the *default* keyword is used, the compiler checks if the minimal requirements specified by the constructs/clauses are met. Although these requirements are checked regardless of this clause, presence of this clause ensures that the compilation will not fail upon not meeting the specified requirements; rather, the OpenMP pragma will be ignored. When the *concept-name* form is used, the adjoining list is checked against this concept. The OpenMP directive is *enabled* if and only if the *types* of all elements in the list model the given concept. Otherwise, the OpenMP construct in question is discarded. When a *cond-expr* is used, the construct is *enabled* only when the condition evaluates to true. Note that this condition does not have to be evaluatable at compile time. The requires clause is intended to be a strict add-on to the requirements specified for the reduction and for construct extensions. Hence, it would not be allowed to undermine the minimal requirements laid down by either of these extensions.

Use case. Using the requires clause, we can express the ultimate generic version of accumulate in ConceptC++ syntax as follows:

```
template<typename InIter, typename T, typename Func>
requires InputIterator<InIter> && BinaryFunction<Func>
        && CopyConstructible<T> && Assignable<T>
T accumulate (InIter first, InIter last, T init, Func f) {
#pragma omp parallel for reduction(f:init) requires(default) requires(Commutative:f) \
                                requires((last−first) > 100)
    for (;first!=last;++first) init = f (init, *first);
}
```

Here, the parallel for construct is enabled if and only if all three requires clauses evaluate to be true.Note that the final requires clause (i.e.,(last−first) > 100) is a runtime condition (effected by injection of appropriate conditional code).

Because this definition of accumulate is so concise, its significance may not be obvious. However, it is quite remarkable. With a single directive, we have enabled parallelization of a generic algorithm, without changing the algorithm itself. The resulting algorithm can be used exactly as before without the directive, i.e., with any types that model the required concepts. Users of accumulate enjoy the benefits of automatic parallelization. If accumulate is invoked with types that model the concepts specified in the OpenMP directive, the algorithm will be parallelized.

4   Prototype Implementation

Implementation of our proposed extensions is currently being carried out with the ConceptGCC compiler, a prototype compiler that implements the ConceptC++ proposal. Based on the GCC 4.3 branch, ConceptC++ therefore also includes OpenMP support from the GOMP project.

We have completed implementation of the for loop construct and reduction clause extensions. Much of the implementation revolves around source code transformation techniques that ensue once the concept requirements of the constructs are satisfied. In case of the reduction clause, we map the reduction operations to primitive OpenMP reductions when the types meet the appropriate requirements thereby avoiding abstraction penalty. For example, a call to accumulate with std::plus<int> maps to + with ints. In cases where we cannot map it to primitive operations, the loop is transformed to perform the reductions manually (for example, string concatenation).

Performance is a paramount concern of generic programming. The performance goal is that generic algorithms should exhibit exactly the same performance as their non-generic counterparts. For example, accumulate invoked with pointer types should have the same performance as a raw for loop with those same pointer types. In the parallel case, we similarly require that a parallelized generic algorithm exhibit the same performance as a parallelized non-generic algorithm. Experimental results with our prototype show that no overhead is introduced by the proposed extensions.[1]

5   Conclusion

We have proposed extensions to OpenMP to provide first class support for parallelization of generic libaries. By providing only three extensions (for, reduction, and requires), we were able to provide this support while simultaneously staying true to the principles of generic programming and OpenMP. As proposed, parallelizing generic algorithms requires only the introduction of OpenMP directives, yet the algorithms so parallelized fully retain their genericity. This capability will be an important one as library developers and application developers seek to take advantage of ubiquitous parallelism. Moreover, by relating parallelizability to concepts, new opportunities for model checking and debugging are now also open.

An important area for future work has to do with the nature of parallel algorithms themselves. The proposed approach is effective for those algorithms for which the sequential and parallel variants admit the same expression. However, in some cases, the parallel version of an algorithm differs significantly from the sequential version. In this case, overloading will still provide a single functional interface, but specialization may be required for various implementations. Characterizing the manner in which generic algorithms must be modified for parallelization will allow principled construction of parallel generic libraries using OpenMP.

References

1. Asanovic, K., et al.: The landscape of parallel computing research: A view from berkeley. Technical Report UCB/EECS-2006-183, EECS Department, University of California, Berkeley (December 2006)
2. Austern, M.H., Towle, R.A., Stepanov, A.A.: Range partition adaptors: a mechanism for parallelizing STL. SIGAPP Appl. Comput. Rev. 4(1), 5–6 (1996)
3. OpenMP Architecture Review Board. OpenMP Application Program Interface, version 3.0 draft (October 2007)
4. Garcia, R., Järvi, J., Lumsdaine, A., Siek, J., Willcock, J.: An extended comparative study of language support for generic programming. Journal of Functional Programming 17(2), 145–205 (2007)

---

[1] We do not currently include detailed results due to page limitations.

5. Gregor, D., Järvi, J., Siek, J., Stroustrup, B., Dos Reis, G., Lumsdaine, A.: Concepts: Linguistic support for generic programming in C++. In: Proceedings of the 2006 ACM SIGPLAN conference on Object-oriented programming, systems, languages, and applications (OOPSLA 2006), October 2006, pp. 291–310. ACM Press, New York (2006)
6. Gregor, D., Stroustrup, B.: Proposed wording for concepts (revision 3). Technical Report N2421=07-0281, ISO/IEC JTC 1, Information Technology, Subcommittee SC 22, Programming Language C++ (October 2007)
7. Kambadur, P., Gregor, D., Lumsdaine, A.: Parallelization of generic libraries based on type properties. In: Proceedings of the 7th International Conference on Computational Science, Beijing, China, May 2007. LNCS, Springer, Heidelberg (2007)
8. Kambadur, P., Gregor, D., Lumsdaine, A., Dharurkar, A.: Modernizing the C++ interface to MPI. In: Proceedings of the 13th European PVM/MPI Users Group Meeting, Bonn, Germany, September 2006. LNCS, pp. 266–274. Springer, Heidelberg (2006)
9. Musser, D.R., Stepanov, A.A.: Algorithm-oriented generic libraries. Softw. Pract. Exper. 24(7), 623–642 (1994)
10. Plauger, P.J.: The Standard Template Library (STL). C/C++ Users Journal 13(12), 10–20 (1995)
11. Reinders, J.: Intel Threading Building Blocks. O'Reilly, Sebastopol (2007)
12. Shah, S., Haab, G., Petersen, P., Throop, J.: Flexible control structures for parallelism in OpenMP. Concurrency - Practice and Experience 12(12), 1219–1239 (2000)
13. Suess, M., Leopold, C.: A User's Experience with Parallel Sorting and OpenMP. In: Proceedings of Sixth European Workshop on OpenMP - EWOMP 2004. LNCS, pp. 23–28. Springer, Heidelberg (2004)
14. Suess, M., Leopold, C.: Problems, workarounds and possible solutions implementing the singleton pattern with C++ and OpenMP. In: International Workshop on OpenMP (2007)
15. Sutter, H.: The free lunch is over: A fundamental turn toward concurrency in software. Dr. Dobb's Journal (January 2005)
16. Terboven, C., an Mey, D.: OpenMP and C++. In: International Workshop on OpenMP (2006)

Streams: Emerging from a Shared Memory Model

Benedict R. Gaster

ClearSpeed Technology Plc
S/W Architecture Group
brg@clearspeed.com

Abstract. To date OpenMP has been considered the work horse for data parallelism and more recently task level parallelism. The model has been one of shared memory working in parallel on arrays of a uniform nature, but many applications do not meet these often restrictive access patterns. With the development of accelerators on the one hand and moving beyond the node to the cluster on the other, OpenMP's shared memory approach does not easily capture the complex memory hierarchies found in these heterogeneous systems.

Streams provide a natural approach to coupling data with its corresponding access patterns. Data within a stream can be easily and efficiently distributed across complex memory hierarchies, while retaining a shared memory point of view for the application programmer.

In this paper we present a modest extension to OpenMP to support data partitioning and streaming. Rather than add numerous new directives our approach is to utilize exiting streaming technology and extend OpenMP simply to control streams in the context of threading. The integration of streams allows the programmer to easily connect distinct compute components together in an efficient manner, supporting both, the conventional shared memory model of OpenMP and also the transparent integration of local non-shared memory.

1 Introduction

OpenMP's shared memory model is one of its strongest points, providing a simple view of memory for the programmer. However, to increase memory bandwidth and reduce memory contention many of today's processors have complex memory hierarchies that do not directly fit this model. For example, both IBM's Cell [1] and ClearSpeed's CSX [2] processors have small single cycle memories attached to local processing elements. These memories are not memory mapped into the larger outer memory system and thus are not shared in the conventional sense, rather data is moved to and from shared memory via DMA transfers. These memories are a problem for the OpenMP programmer as there is no easy why to describe the connection between objects in shared memory and corresponding objects in local memory. Moreover, these memories are often small in size, at most in the region of hundreds of kilobytes, and it is often impossible to keep the complete data set in memory at any given time. Data needs to be "streamed".

R. Eigenmann and B.R. de Supinski (Eds.): IWOMP 2008, LNCS 5004, pp. 134–145, 2008.

In this paper we describe a primitive streaming API that when embedded in a modest extension of OpenMP provides for a powerful alternative to the conventional array based parallel programming model. In particular, it is possible to express complicated non-uniform access patterns for streams that are not easily expressed in OpenMP as is. Streams [3,4,5,6] are best described as a declarative interface to conventional C/C++ style data arrays, that provide for a parallel evaluation semantics, standard and user defined scatter/gather access, and a small set of combinators for writing stream computations. Streams are defined and referenced from anywhere within an OpenMP program, with data pushed and pulled across thread boundaries as specified by the user.

We take the rather unconventional approach of assuming a new basic type, i.e. streams, in the base language. However, it should be noted that we are not proposing extending the base language itself, rather we assume that a streaming API is provided as a library in the particular language of choice. There are many examples of streaming APIs and their implementations are well understood and given this it does not seem unreasonable to build upon these developments [3,4,5,6]. The advantage is that given a non-parallel program, written using streams, it is a natural process to add (extended) OpenMP directives to parallelize for a multi-core environment. This scheme is analogous to that of adding OpenMP directives today in the context of arrays.

In this paper we make the following contributions:

- We describe a modest extension to OpenMP's programming model based on the notion of streams. This model provides an alternative to the conventional array approach, conceptually extending OpenMP's memory model to work in the context of non-shared memory.
- We are implementing a prototype of OpenMP extended with streams for an IA-32 and CSX accelerator based system and we outline its current status.
- We report our experience of using the proposed tasking model for OpenMP with streams, highlighting its natural use in this context.

The remaining sections of this paper are as follows: Section 2 discusses related work; Section 3 introduces, by way of example, how streams can be utilized in OpenMP; Section 4 details the streaming API and the extensions to the OpenMP directives; Section 5 outlines our implementation with some early performance results; and finally Section 6 concludes with a discussion on possible directions for the future.

2   Related Work

To our knowledge there has been very little previously published work on extending OpenMP with a notion of streams. One exception to this is the ACOTES project defining a programming model for streams with a corresponding abstract streaming machine [7]. Carpenter et al. propose a new streaming environment consisting of a Stream Programming Model (SPM), implemented as an annotated version of the C programming language, and an Abstract Streaming Machine (ASM), implemented as a cost-model simulator. Their approach is similar to what is proposed in this paper, although we present extensions to both the data parallel and tasking features of OpenMP while they consider only the tasking aspects. To date they have not implemented their approach and work in this area is necessary to better understand how useful it will be in practice.

There exist a number of proposals for mapping the shared memory model of OpenMP to a distributed setting which is closely related [8,9,10]. These approaches have some important advantages over explicit distributed programming models, such as MPI, including they are conservative extensions to OpenMP and retain the shared memory model. However, the drawback is that they retain this model at the expense of restricting control of data partitioning and movement to the system, thus constraining expressiveness. Providing streams as a first class data abstraction retains OpenMP's shared memory model while exposing control of both data partitioning and movement within a distributed memory setting.

Eichenberger et al. use a software cache to abstract the Cell's SPEs local memory, for both code and data, providing a transparent, shared memory, view of memory [11]. The advantage of this approach is no new datatypes need to be introduced and thus no unnecessary source code changes. What is less clear is how well this approach works in the light of more complex memory layouts which may include many levels of indirection. The overhead of maintaining a software cache in this context could easily dwarf any benefits of such an approach.

The streaming library given in Section 4 is a variant of the streams of Open Accelerator [3]. Open Accelerator is a programming environment that supports accelerator specific code with the integration of streaming, allowing the programmer to easily and efficiently connect distinct components of a system. Open Accelerator is itself orthogonal to OpenMP but Gaster et al. show by superimposing Open Accelerator the resulting system provides a powerful SPMD data parallel and tasking programming abstraction for accelerated systems. A key difference between Open Accelerator and this paper is that streams become a data abstraction that OpenMP builds parallelism upon and are integrated into the language, requiring no additional features or support.

Finally, our work on streams has continually built on the ideas of Brook [6] and StreamIt [5], which provide the stream processor abstraction. While these languages do not fit directly with an imperative programming model it is clear that they provide a wealth of resources for the development of streaming techniques in such a context.

3   Overview

When considering adding a new feature to OpenMP, the place to start, at least one would believe, is the current set of existing directives and possibly some new ones. This was our initial approach when looking at streams for OpenMP and we developed a set of additional directives for working with streams along with some new runtime functions and extensions to existing directives. As an example of this approach consider code to perform a sum of squares for an input array, a, given in Figure 1.

```
#define CHUNK_SIZE 5

double sumsq(double a[], int size) {
    double msum = 0.; int i, n;

#if _OPENMP
    omp_stream_set_chunk_size(CHUNK_SIZE);
#endif

#pragma omp stream create(s, a, size, sizeof(float),
                    LINEAR_FORWARD)

#pragma omp for reduction(+:msum) connect(s:size)
    for (i = 0; i < size; i++) {
        double elem;
#pragma omp stream read(s, elem)
        elem = a[i];

        msum += elem * elem;
    }

    return msum;
}
```

Fig. 1. Streams API directive extensions: sum of squares

Here a stream *s* is created from array, *a*, with the directive *stream create*, introducing a new stream object into the OpenMP environment. The stream is later consumed by the conventional parallel *for* construct with an additional clause, *connect(s : size)*, telling the system to produce a one-to-many stream channel from the controlling thread to its corresponding children, that will perform the work of the loop body. The loop body itself is implemented as though the function was operating over the original input array and makes no direct reference to *s*. Instead, the *stream read* directive is used to connect the next element of *s* to the private variable *elem* and the following statement is ignored, i.e. elem = a[i]. While straightforward to modify the compiler to ignore the connect statement, this approach reaches beyond the original intention of OpenMP's directives.

From this description of streams one might reasonably ask what is the benefit of this approach over straightforward arrays? Well one important difference is the use of the runtime function to set a stream's "chunk" size and a related reference to *size* in the *connect* clause. A streams *chunk* size corresponds to the number of elements that will be copied to a particular thread when reading from a stream's channel. The actual pattern for reading *chunk* number of elements is captured in the final argument to the *stream create* directive; in the above example a simple linear-forward pattern is assumed. A consequence of chunks is that each thread must maintain a private buffer of *chunk* size that when empty

```
double sumsq(double a[], int size) {
  double msum = 0; int i;

  Stream * s = Stream.create(a, size, CHUNK_SIZE, LINEAR_FORWARD);

#pragma omp parallel reduction(+:msum) connect(s)
  {
    while (!s->endOfStream(s)) {
      double x = s->getNextElement();
      if (s->streamActive()) {
        msum += x * x;
      }
    }
  }

  s->destroy();

  return msum;
}
```

Fig. 2. Streams API part of base language: sum of squares

is refilled with *chunk* or less elements by making a read request to the source stream. The observant reader may now be asking but if *chunk* $\neq$ 1 then the loop will be parallelized across *size* threads, but only the first *size/chunk* threads will actually read elements of the stream and even worse the active threads will process only the initial chunk element. Fortunately, this is easily over come by requiring that the user link the controlling bounds variable, *size* in this case, when connecting a stream, which is then used to control the number of iterations.

It should be clear that a compiler is free to implement OpenMP with or without the streaming extensions, assuming it at least parses the *connect* clause, while preserving the semantics of the program. This is, of course, in keeping with the original design ethos of OpenMP, allowing both incremental parallelization and programs execute correctly, albeit often with a slower execution time, if the program is not parallelized at all. The problem with such an approach is that OpenMP was not designed on top of a language with native stream support and adding them explicitly to OpenMP is not only clumsy but it feels like using a bulldozer to crush a nut. For C++ and Fortran arrays are the basic type on top of which OpenMP builds parallelism but with the introduction of task level parallelism this no longer needs to be the case and it should be possible to introduce other parallel data types, including streams. This then leads us to the question: what if streams were provided as a basic datatype?

If streams are provided as an abstract type with a corresponding API, then application programmers could write stream programs with or without OpenMP directives. This meets our goal that any OpenMP program can be compiled and executed correctly; even in the presence of a compiler that does not support OpenMP. Such an approach does not preclude the need to add streams to OpenMP, rather

```
    while (!s->endOfStream(s))
#pragma omp task capturevalue(s)
    {
      double x = s->getNextElement();
      if (s->streamActive()) {
        msum += x * x;
      }
    }
```

Fig. 3. Combining streams and tasks

it provides the foundations for a more modest extension. Figure 2 shows how the sum of squares example might be written with this approach.

Intuitively, a stream s is created from an input array a and the parallel regions creates a team of threads, as is normal, with the only difference that each thread t_i having a corresponding private stream s_i that is "connected" to the input stream s. The behavior is similar to that of a variable marked as private, except instead of copying s locally within the thread's stack, a connection is made between the private s_i and the stream s of the shared enclosing scope. On entering the loop, for a particular thread, the private stream requests a new chunk and if data greater than zero and less than or equal to the number of elements in a chunk is received then the request returns the first of the elements for the call *getNextElement*. If no elements at all are received, then the call to *streamActive* returns false and the sum is not executed and finally the loop will terminate, otherwise the element is squared and added into the running total and the process repeats.

At first viewing the call to *streamActive* may seem unnecessary but there are actually two reasons for its inclusion. Firstly, what happens in the time between the call to *endOfStream* and *getNextElement*? In fact anything and in particular as we are running in parallel the stream may get locked and read by some other thread in the farm and thus leave the call undefined. Secondly, if we are to support SIMD or predicated processors, where conditionals do not necessarily imply control flow, then *streamActive* can provide functionality to disable or enable particular processing elements.

Of course, in the case when the amount of work on each element of a chunk is large the while loop itself can be parallelized with task parallelism. For example, the loop of Figure 2 might be expressed as in Figure 3.

4   Extending OpenMP with Streams

In this section we introduce a streaming API and a small addition to OpenMP's parallel fork-join and producer-consumer threading models.

First we define streams as declarative representations of more conventional random access C++/Fortran arrays. Random access to streams is not allowed, and consequently no index operator exists; instead the user can define a gather/

```
template<typename T> class Stream {
public:
    // Creating and destroying streams
    static Stream<T> * create(T * p, int in_stream_size,
                              int chunk_size,
                              tuple<StreamAccess, TargetISA> *t);
    static Stream<T> * create(Stream<T> *);
    void destroy(void);
    // Reading and writing streams
    T getNextElement(void);
    T * getNextChunk(void);
    void writeNextElement(T&);
    // Stream info
    void flush(void);
    bool endOfStream(void);
    bool streamActive(void);
    int numChunks(void);
    int chunkSize(void);
    // Stream reduction
    template<typename U> U reduce(function<U (U, U)>);
private:
    // constructor, destructor, etc...
};
```

Fig. 4. Steaming API (C++ variant)

scatter style access, using either a set of statically defined patterns or dynamically using streams themselves.

The streaming interface is split into four components: types for streams and access patterns; functions for creating and destroying streams; functions to read and write streams; and functions returning stream characteristics. A stream is created with an array forming the data stream, the size of input, and a list of access patterns. It may not at first be obvious why a list of access patterns is required, rather than just a single value. A single access pattern works fine in the case that a particular stream is destined for a single source but consider the case when a particular stream is distributed across a number of different ISAs, e.g. accelerator cores, which may themselves spread the received data across any number of internal cores. It becomes necessary to associate a particular target ISA (*TargetISA*) with a corresponding stream access, allowing the stream implementation to know at which level a distribution is to be applied.

The streaming functions should be reasonably self-explanatory and the complete set is given in Figure 4.[1] A full description of the streaming API is beyond the scope of this paper and the interested reader is pointed to Gaster et al. for a detailed presentation [3].

For OpenMP itself Figures 5(a) and 5(b) show the extensions necessary for the parallel fork-join and producer consumer directives. As discussed in the previous

---

[1] For simplicity we use C++ as our base language.

#pragma omp parallel [clause[[,]clause]...]   #pragma omp task [clause[[,]clause] ...]
structured-block   structured-block

where clause can be one of
firstprivate(list)
private(list)
shared(list)
reduction(operator:list)
num_threads(integer-expression)
connect(list)

where clause can be one of
captureprivate(list)
private(list)
shared(list)
switch
connect(list)

(a) Fork-join threading model   (b) Producer-consumer threading model

Fig. 5. Streaming extensions to OpenMP pragmas

section this is reduced to a single additional clause connect, capturing the notion that a stream defined at an outer scope is to be joined (one-to-many) to the parallel region's gang of threads. The stream creation method

```
static Stream<T> * create (Stream<T> *);
```

is provided for this and *joins* a stream to the calling region, which in this case will be a thread in the parallel region. The resulting stream is then used in place of the referenced stream within the structured block. On exit from the parallel region the created stream(s) must be destroyed. For output streams this will cause the streams to be flushed and data will be moved to the corresponding stream of the outer scope.

An important consideration when describing a new API for OpenMP must be how easily it can be expressed in C, C++, and FORTRAN. We choose to specify the streaming library in C++ as the parametric polymorphism provided by templates leads to a simple definition. However, while maybe not as compelling when expressed in C or FORTRAN it is straightforward. With careful use of the preprocessor it is possible to generate much of the boiler-plate code necessary for parameterized stream types while retaining most of the generic approach offered by C++. For example, the code in Figure 6 implements a C macro, $STREAMING\_TYPE(typ)$, that when instantiated generates the set of steaming functions for streams of element type typ. While at times not the most elegant of approaches, it does mean that it is possible to retain type safety and makes it applicable in the context of FORTRAN.

5   Evaluation

The streaming environment described in this paper has already been implemented within ClearSpeed for developing applications for an IA-32 system with any number of CSX accelerators. The implementation is factored into two parts: a source-to-source compiler, based on the Barcelona Supercomputing Center's Mercurium compiler [12]; and runtime components for both the IA-32 and CSX.

```
#define STREAMING_TYPE( typ )                                          \
  static typ * get_next_ ## typ ## _chunk (stream s) {                 \
    return ((typ *) get_next_chunk(s));                                \
  }                                                                    \
  static typ get_next_ ## typ ## _element (stream s) {                 \
    return (* ((typ *) (host_get_next_element_p (s))));                \
  }                                                                    \
  static void write_next_ ## typ ##                                    \
                              _element(stream s, typ x) {  \
    write_next_element_p (s,(char *) &(x));                            \
  }                                                                    \
  static void                                                          \
  init_ ## typ ## _stream (stream s,                                   \
            typ * p,                                                   \
            typ * buf0,                                                \
            typ * buf1,                                                \
            int in_stream_size,                                        \
            int out_buf_size,                                          \
            stream_type t) {                                           \
    init_stream (s, (char *) p, (char *) buf0, (char *) buf1,\
              in_stream_size, out_buf_size, sizeof(typ), t); \
  }

STREAMING_TYPE( int );
STREAMING_TYPE( float );
STREAMING_TYPE( double );
```

Fig. 6. Using C's preprocessor to generate a stream API

The streaming API itself is a standard ClearSpeed product for both IA-32 and CSX and required no modifications. The OpenMP compiler expects a single source input, expressed in C++ with OpenMP SIMD accelerator regions [13], that is processed to produce corresponding IA-32 (C++) and CSX (C^n [14]) code, compiled by respective compilers.

For this paper we have evaluated our implementation against a small number of representative benchmarks for performance evaluation. Rather than considering the performance ratio between a native IA-32 implementation and a CSX accelerated system—it is easy to show performance improvements for applications with CSX assistance [2,13]—we considered differences for implementations in OpenMP with and without streams.

The performance figures for each of the selected benchmarks are given in Figure 7[2]. The results themselves are of an early nature but they are very encouraging as is evident, in particular, from the results for the FFT (performing 10,0000 1k and 2k 2D FFTs), sum (sum of squares of one million doubles), and

---

[2] All benchmarks were compiled with GCC 4.1, optimization level -03, and Clear-Speed's latest CSX SDK (3.0).

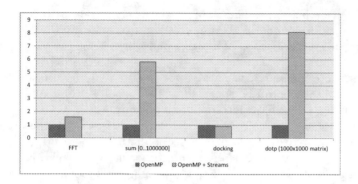

Fig. 7. Performance for OpenMP + Streams vs OpenMP codes

dotp (dot product of 1k vector) which all show a performance improvement over the OpenMP versions, the latter showing a 5x and 7x speedup.

6   Conclusion

In this paper we have presented an approach to extending OpenMP's shared memory model for systems where not all memory is shared. Taking the (possibly) surprising step of assuming a primitive notion for streams it is possible to extend OpenMP in a modest fashion. The programmer is provided with a powerful parallel programming abstraction with the ability to describe irregular data access across distributed memory hierarchies.

In practice the streaming extensions are small and maintain the semantics of existing programs while providing a natural approach to data partitioning not present in OpenMP as it stands. Streams themselves provide a natural parallel programming abstraction and it seems only sensible to consider their application in a parallel programming language such as OpenMP.

With the introduction of streams as alternative containers to arrays, task level parallelism may (often) be more convenient than the conventional data parallel constructs. This is probably due, in part at least, to the fact that OpenMP's original parallel constructs were designed with large scale data-parallelism in mind. With the introduction of task level parallelism it is possible to consider irregular data access that fits well with the produce-consumer style model that arises naturally when working with streams.

6.1   Future Work

Multimedia codecs, such a MEG2 and software radio, show a high amount of data-parallelism and initially seem like a good fit for the data parallel constructs of OpenMP. The problem is that often fine-grained control flow and data communication is required that makes simply loop parallelization difficult and task parallelism finds a better fit. These applications also fit well with gather/scatter and compute style semantics and an interesting area of future work is to develop implementations for a selection of multimedia codecs using the model described in this paper.

As the streams presented in this paper are treated as declarative objects and their computations can be specified using a small set of combinators. It is possible, in many cases, for the compiler to optimize the generation of possible intermediate streams statically. A more detailed discussion of these and other stream optimizations is outside of the scope of this paper and the interested reader is pointed to work on Open Accelerator for more information [3]. To date we have not evaluated the use of these kinds of optimizations in the presence of OpenMP but believe this to be an interesting avenue for future work.

The shared memory model of OpenMP is known to have problems when scaling to simultaneous multi-threaded (SMT) [15] processors. In particular, when OpenMP applications are executed on SMT architectures many different forms of interference between threads has been reported [16]. While not the motivation for the work described in this paper we believe that streams may provide an approach to parallel data access that avoids many of the data interference issues on SMT systems and this is an important area of future work.

Finally, it possible that an implementation of OpenMP built on top of a streaming API could implicitly connect streams to parallel regions without the need for the *connect* clause at all. One problem with this approach comes when considering extending the *connect* clause to capture information on intermediate stream production, which in practice could be optimized away, see Gaster et al. [3] for an example. In this case an explicit stream connection provides vital static information that may otherwise be hidden from the compiler.

References

1. Pham, D., Asano, S., Bolliger, M., Day, M.N., Hofstee, H.P., Johns, C., Kahle, J., Kameyama, A., Keaty, J., Masubuchi, Y., Riley, M., Shippy, D., Stasiak, D., Suzuoki, M., Wang, M., Warnock, J., Weitzel, S., Wendel, D., Yamazaki, T., Yazawa, K.: The design and implementation of a first-generation CELL processor. In: Solid-State Circuits Conference, 2005. Digest of Technical Papers. ISSCC, vol. 1, pp. 184–592. IEEE International, Los Alamitos (2005)
2. ClearSpeed Technology Plc: White paper: CSX Processor Architecture (2004)
3. Gaster, B.R., Lacey, D., Sumner, B.: Open Accelerator: programming at the edges (submitted 2007)
4. Tarditi, D., Puri, S., Oglesby, J.: Accelerator: using data parallelism to program GPUs for general-purpose uses. In: ASPLOS-XII: Proceedings of the 12th international conference on Architectural support for programming languages and operating systems, pp. 325–335. ACM Press, New York (2006)
5. Thies, W., Karczmarek, M., Amarasinghe, S.P.: StreamIt a language for streaming applications. Computational Complexity, 179–196 (2002)
6. Buck, I.: Brook: A Streaming Programming Language. Stanford University (2001)

7. Carpenter, P., Rdenas, D., Martorell, X., Ramrez, A., Ayguad, E.: A Streaming Machine Description and Programming Model. In: Vassiliadis, S., Bereković, M., Hämäläinen, T.D. (eds.) SAMOS 2007. LNCS, vol. 4599, pp. 107–116. Springer, Heidelberg (2007)
8. Huang, L., Chapman, B., Liu, Z.: Towards a more efficient implementation of OpenMP for clusters via translation to Global Arrays. Parallel Comput. 31(10-12), 1114–1139 (2005)
9. Basumallik, A., Min, S.J., Eigenmann, R.: Programming distributed memory systems using openmp. In: 12th international workshop on High-Level Parallel Programming Models and Supportive Environments (2007)
10. Sato, M., Harada, H., Hasegawa, A., Ishikawa, Y.: Cluster-enabled OpenMP: An OpenMP compiler for the scash software distributed shared memory system. Sci. Program 9(2,3), 123–130 (2001)
11. Eichenberger, A.E., O'Brien, J.K., O'Brien, K.M., Wu, P., Chen, T., Oden, P.H., Prener, D.A., Shepherd, J.C., So, B., Sura, Z., Wang, A., Zhang, T., Zhao, P., Gschwind, M.K., Archambault, R., Gao, Y., Koo, R.: Using advanced compiler technology to exploit the performance of the Cell Broadband EngineTM architecture. IBM Syst. J. 45(1), 59–84 (2006)
12. Barcelona Supercomputing Center: The NANOS Environment
13. Bradley, C., Gaster, B.R.: Exploiting loop-level parallelism for SIMD arrays using OpenMP. In: International Workshop on OpenMP (2007)
14. Lokhmotov, A., Gaster, B.R., Mycroft, A., Stuttard, D., Hickey, N.: Revisiting SIMD programming. In: 20th InternationalWorkshop on Languages and Compilers for Parallel Computing (2007)
15. Curtis-Maury, M., Ding, X., Antonopoulos, C.D., Nikolopoulos, D.S.: An evaluation of OpenMP on current and emerging multithreaded/multicore processors. In: First International Workshop on OpenMP (2005)
16. Zhang, Y., Burcea, M., Cheng, V., Ho, R., Voss, M.: An adaptive OpenMP loop scheduler for Hyperthreaded SMPs. In: International Conference on Parallel and Distributed Computing Systems, pp. 256–263 (2004)

On Multi-threaded Satisfiability Solving with OpenMP

Pascal Vander-Swalmen, Gilles Dequen, and Michaël Krajecki

MIS, CReSTIC
{gilles.dequen,pascal.vander-swalmen}@u-picardie.fr,
michael.krajecki@univ-reims.fr

Abstract. The boolean satisfiability problem SAT is a well-known NP-Complete problem, which is widely studied because of its conceptual simplicity. Nowadays the number of existing parallel SAT solvers is quite small. Furthermore, they are generally designed for large clusters using the message passing paradigm. These solvers are coarse grained application since they divide the search-tree among the processors avoiding communication and synchronization. In this paper MTSS, for *Multi Threaded Sat Solver*, is introduced. It is a fine grain parallel SAT solver, in shared memory. It defines a rich thread in charge of the search-tree evaluation and a set of poor threads that will help the rich one by simplifying the opened node. MTSS is well designed for multi-core CPU since it reduces the memory allocation during the search.

Keywords: combinatorial optimization, satisfiability, DLL, collaborative, OPENMP, parallel.

1 Introduction

The boolean satisfiability problem (short for SAT) is a well-known NP-Complete problem[1]. During the last decade, the interest in studying SAT has grown significantly because of its conceptual simplicity and its ability to express a large set of various problems. Nowadays, it remains a central problem in artificial intelligence, logic and computational complexity theory. Thus, it was recently used as a guide to show the convergence between combinatorial optimization and the statistical physics of disordered systems and to propose a new class of algorithms [2]. Within a more practical framework, a lot of works highlight SAT implications in "real world" problems as diverse as Planning [3], Model Checking [4], Cryptography [5], VLSI design, ... In recent years several improvements dedicated, on the one hand to the original backtrack-search DLL procedure [6], and on the over hand to the logical simplification techniques [7] have allowed SAT solvers to be very efficient in solving huge problems from industrial areas [8][1].

In spite of the actual trend in processor development which is from single-core to multi-core CPU, there is few parallel solving approaches dedicated to the SAT

---

[1] http://www.satcompetition.org

R. Eigenmann and B.R. de Supinski (Eds.): IWOMP 2008, LNCS 5004, pp. 146–157, 2008.
© Springer-Verlag Berlin Heidelberg 2008

problem and more generally to the combinatorial problems solving. In fact, the parallel solvers available in the literature are generally dedicated to the message passing paradigm. Even if some CSP (Constraint Satisfaction Problems) solvers in shared memory exist [9], they mainly distribute the search-tree among the available processors. In this paper, we present a collaborative approach dedicated to solve combinatorial problems on shared memory architecture using the OPENMP Application Program Interface. In other words, only one thread, named the *rich thread*, will branch the search-tree while the others, named the *poor threads* will try to simplify the opened, not visited by the rich thread, branches near it. The key advantage of this solution is the data locality involved by this approach which is interesting when considering multi-core CPU.

This paper is organized as follows. In section 2 we briefly describe the SAT problem and provides an overview of the main techniques used to efficiently solve it within a sequential framework. Section 3 presents the different works dedicated to the parallel implementation of the DLL procedure. We describe our approach using the OPENMP API in the section 4. Finally, we provide some experimental results in section 5

2   Preliminaries

A *CNF-Formula* (Conjunctive Normal Form) $\mathcal{F}$ is a set (interpreted as a conjunction) of *clauses*, where a clause is a set (interpreted as a disjunction) of literals. A literal is a signed propositional variable x or its negation $\bar{x}$. In the following, formula (resp. variable) is used instead of CNF formula (resp. propositional variable). An *interpretation* of $\mathcal{F}$ is an assignment of truth values $\{true, false\}$ to its variables. Given a set of boolean variables and a formula $\mathcal{F}$, the SAT problem is to decide if there exists an interpretation of $\mathcal{F}$ in such a way as to make the formula evaluate to TRUE. When no such an assignment exists, $\mathcal{F}$ is FALSE. In this latter case, we would say that $\mathcal{F}$ is *unsatisfiable*; otherwise it is *satisfiable* and each interpretation satisfying $\mathcal{F}$ is a *solution*.

2.1   The SAT Solving

As mentioned above, SAT is formulated as a decision problem. However, we distinguish two related problems. The first one is *to find an interpretation that satisfies $\mathcal{F}$*. Local search methods [10] are useful in this case. Nevertheless, there is no guarantee that such an algorithm will find a solution even if the problem is known to be solvable. Hence, these approaches are *Incomplete*. The second problem related to SAT is *to provide a proof of the non-existence of a solution of $\mathcal{F}$*. To date, only the enumerative methods which mostly are based on Backtrack-Search are able to prove efficiently the unsatisfiability [11]. Thus, these methods scan the search-space systematically and find a solution to the problem if it exists. If they cannot find a solution, they provide a guarantee that the problem has no solution. Hence, these methods are *complete*.

2.2   The DLL Procedure

In this paper we mainly focus on the parallelization of the complete algorithms where most of them are based on the DLL procedure [6]. The DLL procedure appears in the algorithm 1 if not consider the "then" part of the "Poor Task" label. Assigning a variable $x = true$ (resp. $x = false$) helps to simplify the current formula by satisfying (and then deleting) all the clauses containing x (resp $\bar{x}$) and by contradicting (and then deleting) all the occurrences of $\bar{x}$ (resp. x). This is denoted $\mathcal{F}\backslash x$ in the algorithm 1. A literal is monotonic when its opposite does not belong to $\mathcal{F}$. A *unit clause* is a clause which consists of exactly one literal. Each unit clause will be satisfied by its unique literal (see "Unit Propagation" label in Algorithm 1) unless an empty clause is encountered. The DLL procedure recursively enumerates the search-space by constructing a tree whose paths correspond to variable assignments. At each node of the search-tree, a variable v is chosen and the formula $\mathcal{F}$ is split into two simpler sub-problems $\mathcal{F}\backslash v$ and $\mathcal{F}\backslash\bar{v}$. If at least one of them contains at least one empty clause, DLL backtracks to the nearest (in term of hamming distance) unvisited assignment (see "Backtrack" label). A solution is found when no clause belongs to at least one of them (see "Solution" label).

In order to improve the DLL procedure, the literature proposes some research fields which the mains are:

- *The choice of the splitting variable* (see "Split" label of the algorithm 1) determines the order in which search is executed. It is an essential key to minimize the size of the search-tree. To date, we mainly distinguish splitting policies dedicated for randomly generated problems [11] and for industrial problems [8].
- *The pruning techniques for* DLL are related to all techniques which are able to reduce the domain of the variables. The most known of them is Unit Propagation described above. We refer to the *look-ahead, equivalency reasoning*, and more recently the *clause recording* and *non-chronological backtracking*.
- *The preprocessing of the formula* refers to all techniques which simplifiy the formula before applying DLL. For instance we can find restrictive resolution or hyper-resolution [7] techniques.

3   SAT Parallel Solving

During the last decade, a lot of works to improve the sequential resolution run-time of the SAT problem have been proposed and have allowed SAT solvers to be very efficient in solving formulas from which the size and the solving difficulty increase. Nevertheless, there is to date few parallel solving approaches dedicated to the SAT problem. Moreover, the most of them are dedicated to the message passing paradigm and use the search-space partitionning to assign work to the available threads during the runtime. This often leads to use a master-slave scheme where the most difficult part consists in balancing the workload.

Among the parallel SAT solvers from the literature, we can remark PSATO [12], based on the sequential solver SATO which introduces the important notion of *guiding path*. The guiding path is a dynamic object which represents the partial ordered interpretation of the splitting variables from the root to the current leaf of the search-tree during the backtrack-search process. Thus, it defines disjoint search-spaces respectively assigned to the parallel tasks. The fig. 1(a) provides a sample illustration of it. Thus, each CPU executes the sequential solver on each associated subtree rooted at each node of the guiding path. An important characteristic of the SAT search-space is its unbalanced distribution. Hence, it is hard to predict the time needed to achieve the enumerative process of a branch. Following the same *master-slave* model, //satz [13], which is based on the satz solver, is a parallel distributed solver that uses a dynamic workload balancing. This solver exhibits the *ping-pong phenomenon* [13]. It was essentially dedicated to solve Random k-SAT formula [14]. MiraXT [15] and ySAT [16] use the same scheme as above and integrate the *clause learning* [17] technique in order to share logical informations between tasks ; these two solvers are multi-threaded. GridSAT [18] is a distributed SAT solver based on zChaff [8]. It is especially dedicated to grid computing. Finally, a new approach named JackSAT [19] uses a decomposition and join scheme of the variables set.

4   Our Collaborative Approach

In this paper we propose a new parallel scheme of the DLL procedure. The first of our contribution is to enhance the guiding path notion. Thus, we describe in the fig. 1(b) the notion of *guiding tree* which is a subset of all dangling nodes of the search-tree. This new notion is strongly coupled with the two others concepts: *the rich thread* and *the poor thread*.

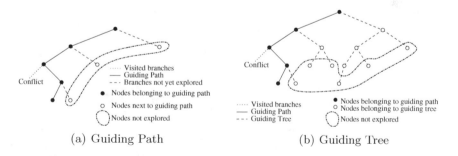

(a) Guiding Path   (b) Guiding Tree

Fig. 1. Guiding Path and Guiding Tree samples

The rich thread is able to decide if a formula is SAT or UNSAT in a finite (and exponential) time. The poor thread is able to provide partial or global information about the formula but without any guarantee (e.g.: unit propagation, choice heuristic of the splitting variable, local search algorithm, look-ahead, preprocessing technique, clause learning ...).

4.1   Rich Thread vs. Poor Thread

MTSS is not a classical partitionning search-tree with similar solving threads. It consists in exactly one rich thread DLL-*like* which is helped with an asynchronous model by one or more poor threads. If none poor contributes to the solving, the rich thread is equivalent to a classical sequential SAT solver. The cooperation between rich and poor threads is done when rich thread backtracks. At this instant, the rich thread is able to use logical or structural informations computed by poor threads. On the other hand, the poor threads are entities which browse the guiding-tree resulting from the rich thread treatment (see fig. 2). They achieve some tasks according to the node status they are watching and initiate themselves their job. Hence, there is no other thread dedicated to assign the jobs (no master/slave model). Moreover, this induces a natural workload balancing.

Algorithm 1. The RICH THREAD (DLL-like) procedure

Require: $\mathcal{F}$: a propositional formula

RICHTHREAD($\mathcal{F}$)

IF $\mathcal{F}$ contains one monotonic literal l. THEN return RICHTHREAD($\mathcal{F}\backslash l$) (**Monotonic Literal**)

IF $\mathcal{F}$ contains one unit clause containing l. THEN return RICHTHREAD($\mathcal{F}\backslash l$) (**Unit Propagation**)

IF $\mathcal{F}$ contains at least one empty clause THEN return FALSE(**Backtrack**)

IF $\mathcal{F}$ is empty THEN return TRUE(**Solution**)

$v \leftarrow$ one unassigned variable of $\mathcal{F}$ (**Split**)

IF DLL ($\mathcal{F}\backslash v$) = TRUE THEN return TRUE

ELSE IF At least one Poor Thread has finished its local calculus on the current node THEN replace current computing context by the Poor Thread's one (**Poor Task**)

ELSE return RICHTHREAD($\mathcal{F}\backslash\bar{v}$))

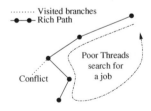

Fig. 2. Poor threads search themselves job (guiding path)

4.2   Poor Tasks

The poor tasks may be numerous. Nevertheless, to date, our solver named MTSS implements two essential poor tasks needed for winning time. The first one is the

Algorithm 2. POORTHREAD procedure

Require: $\mathcal{F}$: a propositional formula
Require: T : a task
  POORTHREAD($\mathcal{F}$, **T**)
  $n \leftarrow$ Root of $\mathcal{F}$-search-tree
  WHILE $\mathcal{F}$ has no solution
    IF T can be applied on n THEN
      Apply T on n
    IF n is the last node of guiding path or threshold reached THEN
      $n \leftarrow$ Root of $\mathcal{F}$-search-tree
    ELSE $n \leftarrow$ next node of the guiding tree from n

---

backtrack and look-ahead node. When the rich thread visits the node below the left branch, a poor thread can computes the right one. Thus, the poor thread pre-computes the formula with the opposite truth value chosen by the rich thread on the left branch. The second task is the splitting variable selection. If the first task is finished on a node, the choice of the next splitting variable can be computed. Thus, if the rich thread must backtrack at this node, the computation of the formula and the heuristic are already done. The fig. 3(a) illustrates these tasks. According to our experimental results, the first task is rarely a win of time cause the time of computation for this task is smaller than the time needed to load the computation context in processor's cache memory which is running the rich thread. The second task takes a long time. Within our practical framework the rich thread does not use the informations computed by poor threads when only the first task is ready. Nevertheless, that can generate useless work from poor threads. To avoid it, we define an empirical threshold value. This value corresponds to the number of splits from the root to the current node. Beyond this value, poor threads don't work (see fig. 3(b)). In this case the rich thread works alone. The default value of this threshold is set to 5% of the total number of variables (with the splitting variable selection used - BSH [11] where the number of splitting variables represents around 10% of total number of variables). Hence, the poor tasks pratically work until the middle depth of the search-tree.

Finally, we define a third poor task with aim to "open" the right nodes of the guiding path and describing the guiding tree. This task can be computed from any node of the guiding tree and computes either left branch or right branch if the left one is in course of computation or finished. After computing the propagation of a literal, it selects the splitting variable. With this third task, the poor threads simultaneously deploy several partial sub-trees (guiding tree) rooted in rich path (the guiding path). This task is shown in fig. 4.

In order to maximize the helpful future work of the rich thread and the future nodes to develop (i.e. the right dangling nodes), the poor thread chooses to first open left branch from a node belonging to the guiding tree.

Table 1 shows the spending time to swap the context informations from poor to rich (see the "rich" columns) and from rich or poor to poor (see the "poor" colums). We clearly can remark that this information exchange cost for rich

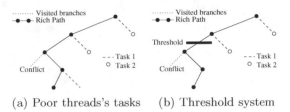

(a) Poor threads's tasks   (b) Threshold system

Fig. 3. Tasks and Threshold

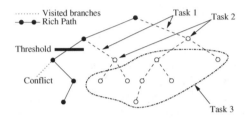

Fig. 4. Example of execution with the third task

thread decreases as the formula size increases. Poor thread spends much more time swapping context than the rich thread one because it always begins a task by copying the current context of a node in its memory, then it computes a task and finally copies the new computed context in the node. The rich thread never changes its context but when it takes informations from poor thread.

Table 1. Context swap cost between one poor thread and one rich thread

Execution Time	0.6s. (240 vars)		9.7s. (300 vars)		40s. (350 vars)	
Thread	rich	poor	rich	poor	rich	poor
# Swap	327	3,764	2,756	53,383	8,188	170,944
% / Time	0.32%	3.63%	0.17%	3.77%	0.13%	3.74%
Time spent (sec.)	0.00192	0.02178	0.01649	0.36569	0.052	1.496

4.3   Memory Management

Our technique leads to irregular memory time and memory space accesses. To insure good performances despite that difficulty, we have implemented a specific memory management. It consists in isolating memory as mentionned in [20] (see fig. 5). The memory allocations are grouped by usage so that each part will be contiguous. Thus, cache-faults (cache-miss not required) are restricted. Some free areas are needed to avoid the allocation of a memory page dedicated to two areas with different usages. So, each kind of memory is isolated:

- Private memory for each thread (arrays of some datas in functions...), moreover each thread memory is isolated too
- Shared and read-only memory (invariant datas as clauses or number of variables...)
- Read & write shared memory (search-tree) and each node is isolated from the others

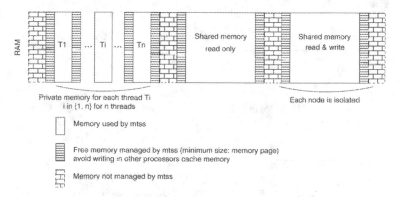

Fig. 5. Memory management

To estimate improvements due to this memory locality, we made some benchmarks shown in fig. 6. We can note a difference between efficiency: with 4 threads, our solver is 78% efficient versus 63% for a modified solver using classical mallocs. This difference of efficiency increases with number of threads.

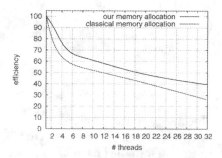

Fig. 6. Efficiency of our approach using two memory managing policies according to the number of threads

To avoid waits among threads, beside some specific locks (final solution lock for example), there are as much locks as branches in the search-tree. Indeed with the aim to have a very thin granularity of parallelism (two threads on the same node), a lot of locks are required. That is the reason for which we put two locks a node.

5   Experimental Results

5.1   Formulas

The formulas tested are random 3-SAT at the pick of difficulty [14] for each size. Each curve is generated from several computations and for different number of threads on ten formulas for each size. The benchmarks were on three different sizes of formulas: 350, 400 and 450 variables. Each curve is the result of a different threshold (fig 7).

The studied formulas are relatively small: the sequential solving time is around 1, 10 and 120 minutes respectively for the formulas of tiny, medium and huge size (see table 2).

Table 2. Informations on sequential running time

#vars	#instances	Sequential Time (s.)		
		min	max	mean
350	10	56.606	105.772	79.3038
400	10	162.554	886.551	517.9526
450	10	1,682.609	5,994.016	3,576.3146

5.2   Protocol

The cluster of SMP used for benchmarks is ROMEO II[2] from the University of Reims. 48 dual Core Itanium 2 (Montecito 4M 1.6Ghz) are dedicated to computation. The cluster is made-up of 6 SMP servers of 8 cores, 1 of 16 cores and the last SMP node offers 32 cores. Each core of the cluster has at least 2 Gbytes of main memory.

MTSS is developed in C language with OpenMP primitives and functions. It has been compiled with the Intel compiler ICC 10.1.

5.3   Results

As mentioned earlier, the objective is to have an efficient SAT solver for multi-core CPU, this is the reason why the experiments are limited to 8 cores.

For each size of the problem, 10 formulas have been generated. Each formula has been two times computed with 1, 2, 4 and 8 processors. So, 240 runs have been conducted to obtain results in fig. 7.

One can observe that MTSS achieves good efficiencies until 8 processors even for the smaller problems (with 350 variables, the efficiency is greater than 60%). When using only 4 cores, the efficiency measured is closed to 80%. The reader may notice that we have a fine grain application since MTSS threads visit more than 1700 nodes per second.

---

[2] http://www.romeo2.fr

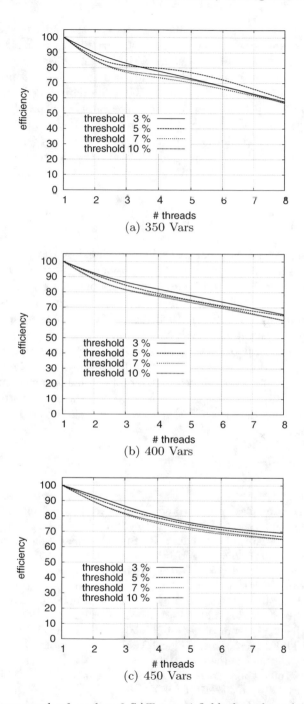

(a) 350 Vars

(b) 400 Vars

(c) 450 Vars

Fig. 7. Efficiency graph of random 3-SAT unsatisfiable formulas solving with a ratio $\frac{\#vars}{\#clauses} = 4.25$ (pick of difficulty) for 350, 400 and 450 variables

Second, when the size of the problem increases, the efficiency is more and more higher. The efficiency observed with 450 variables is near from 70% for 8 processors which is 10 points better than with 350 variables.

Third, one can remind that the threshold used to decide if a poor thread is allowed to treat a node or not, is an arbitrary value. For these experiments the value varies from 3 to 10% of the number of variables. In practice, it is well-known that SAT solvers will in general not go down beyond this value in the search tree. So, when considering a value of 10%, it means that there all the nodes can be visited by the poor threads. By fixing the threshold near from 5%, the efficiency is improved by 5 to 10%.

6   Conclusion

In this paper, we present a new parallel scheme to improve the main state-of-the-art enumerative SAT solving technique and provide an easy way to use and to parallelize the existing sequential deduction techniques. Our approach has been implemented in a new parallel solver named MTSS. The current version of MTSS is efficient, but our solver lacks of maturity and is not yet competitive compared to the best current sequential solvers. Several improvements to speed it up can be done. The number of poor tasks used in MTSS is very limited. Each new poor task is a new hope to improve it and we plan to study some of the existing ones such as the preprocessing techniques, subsumption deduction, clause learning, stochastic local search, . . . Moreover, it could be interesting to study the impact of swapping poor and rich threads status when a rich thread is waiting for a poor computation.

References

1. Cook, S.A.: The Complexity of Theorem Proving Procedures. In: 3rd ACM Symp. on Theory of Computing, Ohio, pp. 151–158 (1971)
2. Braunstein, A., Mézard, M., Zecchina, R.: Survey propagation: An algorithm for satisfiability. Random Struct. Algorithms 27(2), 201–226 (2005)
3. Kautz, H., Selman, B.: Pushing the Envelope: Planning, Propositional Logic and Stochastic Search. In: Proc. of the 30th National Conf. on Artificial Intelligence and the 8th Innovative Applications of Artificial Intelligence Conf., Menlo Park, August 4–8, 1996, pp. 1194–1201. AAAI Press / MIT Press (1996)
4. Biere, A., Heljanko, K., Junttila, T., Latvala, T., Schuppan, V.: Linear encodings of bounded LTL model checking. Logical Methods in Computer Science 2 (2006)
5. Potlapally, N.R., Raghunathan, A., Ravi, S., Jha, N.K., Lee, R.B.: Aiding side-channel attacks on cryptographic software with satisfiability-based analysis. IEEE Trans. VLSI Syst. 15(4), 465–470 (2007)
6. Davis, M., Logemann, G., Loveland, D.: A Machine Program for Theorem-Proving. Journal Association for Computing Machine (5), 394–397 (1962)
7. Bacchus, F., Winter, J.: Effective preprocessing with hyper-resolution and equality reduction (2003)

8. Zhang, L., Madigan, C., Moskewicz, M., Malik, S.: Efficient Conflict Driven Learning in a Boolean Satisfiability Solver. In: Proc. of ICCAD, San Jose (November 2001) (to appear)
9. Habbas, Z., Krajecki, M., Singer, D.: Decomposition techniques for parallel resolution of constraint satisfaction problems in shared memory: a comparative study. Intern. Journal of Computational Science and Engineering (IJCSE) 1(2/3/4), 192–206 (2005)
10. Hoos, H.H., Stützle, T.: Stochastic Local Search: Foundations & Applications (The Morgan Kaufmann Series in Artificial Intelligence). Morgan Kaufmann, San Francisco (2004)
11. Dequen, G., Dubois, O.: An efficient approach to solving random -satproblems. J. Autom. Reasoning 37(4), 261–276 (2006)
12. Zhang, H., Bonacina, M.P., Hsiang, J.: Psato: a distributed propositional prover and its application to quasigroup problems. J. Symb. Comput. 21(4-6), 543–560 (1996)
13. Jurkowiak, B., Li, C.M., Utard, G.: Parallelizing Satz Using Dynamic Workload Balancing. In: Proc. of Workshop on Theory and Application of Satisfiability Testing (Sat 2001), Boston, June 2001, pp. 205–211 (2001)
14. Mitchell, D., Selman, B., Levesque, H.J.: Hard and easy distribution of SAT problems. In: Proc. 10th Nat. Conf. on Artificial Intelligence, pp. 459–465. AAAI, Menlo Park (1992)
15. Lewis, M., Schubert, T., Becker, B.: Multithreaded sat solving. In: ASP-DAC 2007: Proceedings of the 2007 conference on Asia South Pacific design automation, Washington, DC, USA, pp. 926–931. IEEE Computer Society Press, Los Alamitos (2007)
16. Feldman, Y., Dershowitz, N., Hanna, Z.: Parallel multithreaded satisfiability solver: Design and implementation (2004)
17. Silva, J.P.M., Sakallah, K.A.: Grasp a new search algorithm for satisfiability. In: ICCAD 1996: Proc. of the 1996 IEEE/ACM Intern. Conf. on Computer-aided design, Washington, DC, USA, pp. 220–227. IEEE Computer Society, Los Alamitos (1996)
18. Chrabakh, W., Wolski, R.: Gridsat: Design and implementation of a computational grid application. J. Grid Comput. 4(2), 177–193 (2006)
19. Singer, D., Monnet, A.: JaCk-SAT: A New Parallel Scheme to Solve the Satisfiability Problem (SAT) based on Join-and-Check. In: Proc of Parallel Processing and Applied Mathematics, Gdansk (2007)
20. Jaillet, C., Krajecki, M.: Parallel programming with openmp: a new memory allocation model avoiding cache faults. In: Intern. Workshop on OpenMP 2007 (IWOMP2007). Tsinghua University, Beijing, China (June 2007)

Parallelism and Scalability in an Image Processing Application

Morten S. Rasmussen, Matthias B. Stuart, and Sven Karlsson

DTU Informatics
Technical University of Denmark
{msr,ms,ska}@imm.dtu.dk

Abstract. The recent trends in processor architecture show that parallel processing is moving into new areas of computing in the form of many-core desktop processors and multi-processor system-on-chip. This means that parallel processing is required in application areas that traditionally have not used parallel programs. This paper investigates parallelism and scalability of an embedded image processing application. The major challenges faced when parallelizing the application were to extract enough parallelism from the application and to reduce load imbalance. The application has limited immediately available parallelism. It is difficult to further extract parallelism since the application has small data sets and parallelization overhead is relatively high. There is also a fair amount of load imbalance which is made worse by a non-uniform memory latency. Even so, we show that with some tuning relative speedups in excess of 9 on a 16 CPU system can be reached.

Keywords: OpenMP, image processing, parallelization.

1 Introduction

To reach higher performance, processor designers have in the last few decades focused on clock frequency and elaborate designs that can extract implicit parallelism from sequential code. However, presently that approach leads to diminishing returns and high power consumption. As a consequence, vendors have turned their focus to multi-core architectures where several processors are placed on a single silicon chip. Such architectures are inherently explicitly parallel. So far the programming models for multi-core architectures have been very similar to the programming models for shared memory multiprocessors.

Embedded systems follow a similar trend where multi-processor system-on-chip solutions with advanced interconnection networks have been proposed [1,2,3]. Thus, there is a need to explore parallel processing in the context of embedded systems. Thereby new challenges are exposed as embedded systems have requirements different from those of high performance computing systems. In embedded systems, processing time may be more important than processing throughput.

In this paper, we explore an embedded image processing application and we investigate its parallel behavior using OpenMP [4]. In short, our contributions

R. Eigenmann and B.R. de Supinski (Eds.): IWOMP 2008, LNCS 5004, pp. 158–169, 2008.

are: 1) The analysis of an embedded image processing application; 2) A thorough performance evaluation of the parallel properties of the application using OpenMP.

The major challenges faced when parallelizing the application were to extract enough parallelism from the application and to reduce load imbalance. The experimental results show that, with some tuning, relative speedups in excess of 9 on a 16 CPU system can be reached.

The rest of the paper is organized as follows. This section is concluded with a discussion of related work. In Sect. 2, we describe the application and in Sect. 3 we explain how the application has been parallelized. Experimental results are presented in Sect. 4. The paper ends with conclusions in Sect. 5.

1.1   Related Work

Parallelization of image processing algorithms for image classification using OpenMP has previously been presented [5]. The work investigated an algorithm used for identifying forest areas in satellite images. The algorithm is scaled by running individual processing steps in parallel and by splitting the data set into smaller parts. A 64 processor high performance computer was used as test platform to process the 1.2 gigabyte images. Our work differs in that we present experiences with a potential embedded application with images two orders of magnitude smaller, which means that parallelization overhead is more pronounced.

Content-based image retrieval is another application of automated image classification which can benefit significantly from parallelization. Content-based image retrieval allows advanced image database queries based on image content. A database query thereby involve processing every image in the database in order to examine its content. A shared memory parallelization of this application has been presented previously [6]. The application is made parallel by processing individual queries and images involved in each query in parallel. In contrast, we strive to minimize processing latency of a single image by parallelizing the processing of the individual image, which requires more fine grained parallelization.

2   Image Processing Application

In this paper, we are focusing on an image processing application developed at DTU [7] and written in Matlab. The application can be used for many different purposes. For example, it can be used for identifying the species of a *Penicillium* fungus in a petri dish from a multi-spectral image [7]. The information is extracted in the form of scalar values, called *features*, that each describe some aspect of the input image. Features are grouped into *feature sets*, based on extraction method used for the particular features.

The flowchart in Fig. 1 gives an overview of the application. It consists of three major parts of which only two are shown: Pre-processing/mask generation, feature generation based on arithmetic and morphological operations and feature generation based on scale spaces. In this paper, we focus on the pre-processing/mask generation and features from arithmetic operations. The statistical methods for classifying the contents of images are outside the scope of this paper and are described elsewhere [7].

The remaining parts of this section will describe the algorithm in more detail.

2.1   Pre-processing and Mask Generation

Before information can be extracted from the image, two steps have to be run on the input.

The pre-processing step produces a noise-filtered normalized image. First, the pixel-wise average intensity across spectral bands in the multi-spectral input image is found. The mean of the resulting single-channel image is found and subtracted from each pixel in the multi-spectral input. Following this, each pixel is then divided by the standard deviation producing the normalized image. Finally, a 3×3 median filter is used to filter noise, see Fig. 1.

The mask step is used to select the interesting parts of the image, thus its generation varies depending on what information is extracted. For the input images used in this paper, the edge detection is used to find the useful parts of the image. For each pixel, the magnitude of the numerical gradient $|(\frac{\mathrm{d}f}{\mathrm{d}x}, \frac{\mathrm{d}f}{\mathrm{d}y})|$ is calculated where f describes the pixel values as function of coordinates (x, y). The median of the gradient values is found and all pixels whose gradient is greater than or equal to the median are included in the mask. They correspond to interesting areas in the image. The mask can be seen as a bit field where each bit corresponds to a pixel in the image. Each bit indicates if the pixel should be considered or not.

2.2   Arithmetic Feature Extraction

The mask is applied to each spectral band in the input image by discarding all pixels *not* in the mask. Five feature sets are extracted from the masked spectral bands of the input image, using five different arithmetic operations. Two operations take a single band at a time, while the other three operate on all pairs of bands. The two single-operand operations are the no-operation and the base-10 logarithm. The other three operations find the pixel-wise difference, product and quotient of all pairs of bands. Each pair is considered only once, e.g. if $I_a - I_b$ is calculated, $I_b - I_a$ is not.

If the input image has n spectral bands, the operations produce $2n + 3\frac{n(n-1)}{2}$ data sets. The features of each feature set are extracted from the data sets by finding the 1st, 5th, 10th, 30th, 50th, 70th, 90th, 95th and 99th percentiles. Doing so requires the data sets to be sorted individually.

3   Parallelization

We will now discuss the parallelization and the OpenMP implementation of the algorithm described in Sect. 2. The image processing algorithm differs from

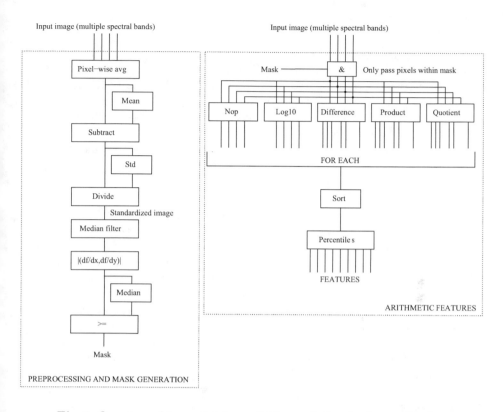

Fig. 1. Overview of immediately available parallelism in the application

traditional high performance computing applications, such as matrix multiplication and physics simulation by having a significantly smaller data set and shorter execution time. Thus, the parallelization overhead can not be neglected.

The algorithm has two main parts as illustrated in Fig. 1. The pre-processing and mask generation part is governed by data dependencies, while the arithmetic feature extraction has parallelism immediately available among the feature sets, but also within the individual sets.

Profiling a sequential implementation of the algorithm revealed that 95% of the execution time is spent in feature extraction. Thus, it is the target for parallelization.

To summarize the task parallelism illustrated in Fig. 1, five independent feature sets are computed, which each produce n or $n(n-1)/2$ data sets for which the features are extracted by finding certain percentiles in the data sets. This means that the processing required for each feature set differs significantly. The feature extraction within each feature set should, in theory, be possible to split into parallel and equally sized workloads. However, non-uniform memory latencies caused by the target architecture may cause the execution time of each such parallel workloads to differ. Scaling properties are discussed in Sect. 3.1 without considering architectural effects which are discussed in Sect. 3.2.

3.1   Scaling Properties

The running times of the feature sets differ by up to a factor of $(n-1)/2$ leading to load imbalance problems if different feature sets are run in parallel. In this paper, we therefore concentrate on extracting parallelism of each individual feature set.

As mentioned earlier, each feature set has n or $n(n-1)/2$ equally sized workloads immediately available, which can be run in parallel. But if n is less than the number of available processors $|P|$, in processor set P, more parallelism must be extracted from these workloads. This is also advantageous to reduce the imbalance slack for the feature sets containing $n(n-1)/2$ workloads, as this may not match a multiple of $|P|$.

Additional parallelism can be extracted by splitting data sets into subsets that can be computed independently and then recombined. Thereby adding an extra nested level of parallelism. The arithmetic operations of all feature sets have no inter-pixel dependencies, which mean that the processing of spectral bands into data sets can be split without creating any subset border synchronization issues. The sorting involved in the percentile calculation, can be done on each subset separately and then merged to recombine the subsets before the percentiles are found. This allows the arithmetic operations to scale further, but with the overhead of merge sorting the subsets. It should be noted that the execution time of sorting each subset decreases by $d \times log(d)$, where d is the number of pixels in the subset, while the time to do the merge sort increases linearly with the number of subsets. This means that the amount of parallel work decreases and the sequential part increases with an increasing number of subsets. Thus, the gain of increasing parallelism is diminishing. In addition, the parallelization overhead, such as spawning threads and synchronization, may be significant at this level as the subsets are small.

The two levels of parallelism within each feature set, among data sets and among subsets, are denoted as l_0 and l_1 respectively. In our implementation, the parallelism at each level s_0 and s_1, can be adjusted independently, though the parallelism at l_0 is limited. The total number of subsets across all data sets w is given by $w = s_0 \times s_1$ and constitutes the total number of workloads in the application. Subset processing time is defined as the wall clock time spent performing arithmetic operations on the parts of the spectral band data that corresponds to the subset and time spent sorting the subset.

In order to avoid load imbalance, s_0 and s_1 should be determined such that w is equal to or slightly less than a $m \times |P|$, where m is a multiple of the number of available processors $|P|$. If w is slightly larger than $m \times P$, only one or a few processors will be involved in processing the last remaining subsets while the majority of processors are idle, causing a large slack. The slack will be reduced by increasing w. But as mentioned earlier, s_0 is limited by n or $n(n-1)/2$ and s_1 is limited by the merge sort overhead, which causes diminishing parallelization gain. Determining s_0 and s_1 is a trade off between load imbalance and parallelization overhead.

3.2   Non-uniform Memory Latency

The discussion in the previous section holds under the assumption that the execution time of equally sized workloads do not differ. This assumption will not hold for architectures with non-uniform memory latencies. Threads running on processors which have long memory latency will have longer subset processing times than threads with short memory latency.

In this application all spectral bands of the image are loaded into memory sequentially and then processed in parallel. Assuming a first touch memory placement policy in a hierarchical memory system, all image data will be located in the part of main memory local to the processor loading in the images, e.g. in the local memory on a Uniboard in a Sun Fire architecture system. A thread running on a processor associated with a different branch of the memory hierarchy, e.g. a processor on a different Uniboard than the one holding the main memory containing the image data, will access all data through the global memory interconnect and therefore have a significant longer memory latency. This is not easily solved through parallel loading of the spectral images due to the fact that the data set processing requires all combinations of spectral bands. Thus, the effective subset processing time depends on the processor.

Combining this effect with the scaling properties, means that even though the total number of subsets w match the number of available processors, linear speedup can not be obtained. Consider a system with $|P|$ processors, where $P_l \subset P$ is the subset of processors having local memory access to the image data and $P_r \subset P$ is the subset of processors having remote memory access to the image data through global memory interconnect. The execution times of a subset on $p_i \in P_l$ and $p_j \in P_r$ are t_l and t_r respectively, where $t_r > t_l$.

In the case of uniform memory latency, where $P = P_l$ and $w = m \times |P_l|$, the total execution time is given by $T = m \times t_l$, neglecting the parallelization overhead. In the non-uniform case where $P = P_l \cup P_r$, T depends on the workload scheduling. Consider the case where w equals the number of processors $|P|$. In this case every processor will process one subset each. Thus the total execution time is given by $T = max(t_l, t_r) = t_r$, if the parallelization overhead is assumed to be negligible. The processors in P_l finish before the processors in P_r, but the final results is not available until all processors have finished processing their subset. In the case where $w = 2 \times |P_l| + |P_r|$, assuming dynamic scheduling, $T = max(2t_l, t_r)$ as the processors in P_l will finish two subsets. If $2t_l > t_r$ the remote memory access of P_r, will not influence T. This is illustrated in Fig. 2. As a consequence of these two cases, resolving load imbalance may not result in the speedup outlined in Sect. 3.1. This applies to scaling both the number of processors and subsets, as these are both parameters that influence the load imbalance. Increasing the number of processors, such that $w = 2 \times |P_l| + |P_{r1}|$ becomes $w = |P_l| + |P_{r2}|$, where $|P_{r2}| = |P_{r1}| + |P_l|$, results in $T = t_r$. Thereby the total execution time reduction is only $2 \times t_l - t_r$, and not t_l.

The effect of load imbalance due to non-uniform memory latency also decreases significantly when w becomes much larger than the number of processors. Then again, the amount of parallelism available in the application may

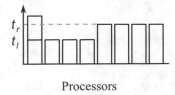

Processors

Fig. 2. Different workload execution times caused by non-uniform memory latency

be limited and comes at a high cost in terms of parallelization overhead. The optimum solution is a trade off between parallelization overhead and load imbalance, where load imbalance is caused both by the algorithm itself, but also the architecture of the target execution platform. It should be noted that this is based on dynamic workload scheduling. Static workload scheduling will perform worse, due to execution time variation among the workloads.

3.3   OpenMP Implementation

The application was originally implemented in Matlab, and then ported to C using standard libraries only, without OpenMP parallelization in mind. It was then modified to meet the requirements for OpenMP parallelization.

In the sequential algorithm implementation, each feature set is implemented as loops, where each iteration performs the arithmetic operation on a spectral band or pair of spectral bands, to form a new data set from which features are extracted. Unary arithmetic operators are applied to each individual spectral band in feature sets 1 and 2. These are implemented by a single loop through all the spectral bands. The feature sets 3, 4 and 5 are based on binary arithmetic operations which are implemented by first generating the list of pairs to be processed, and then process one pair for each iteration in a single loop.

Two different OpenMP versions have been implemented. One of the versions use nested parallelism and the other does not.

The nested version has two levels of parallelism. The first level of parallelism, l_0, consists of the aforementioned loop over data sets. This loop is parallelized using the OpenMP [4] for loop workload sharing construct with dynamic scheduling.

Within each l_0-thread the data set is further split into subsets processed by a loop, which forms the nested parallelism level l_1. One thread is created for every subset. Creating more threads than processors will enable operating system schedulers capable of dynamic thread migration to load balance the processors. However, spawning more threads than processors may also induce a large scheduling overhead in the operating system.

To avoid relying on the operating system thread load balancing capabilities a non-nested version has also been made. To flatten the two levels of parallelism, all $s_1 \times n$ or $s_1 \times n(n-1)/2$ subsets are enumerated and then processed in a single parallel for loop. The number of threads is thereby completely independent of how many subsets the data sets are split into.

To avoid cache and memory effects, the data sets must be merged as soon as all their subsets have been processed. Doing so ensures that as much as possible of the processed data remains in processor caches. This has been implemented by making the thread finishing processing the last subset of a data set merge all the subsets. Experiments have indicated that this can improve performance by 4% to 18%. A consequence of the non-nested version is that the subset processing times are not equal. Thus, dynamic scheduling is used for the workload sharing construct. Although the non-nested version is more complex than the nested, the extra book-keeping code does not negatively influence the execution time.

4 Results and Discussion

This section presents results obtained by running the nested and non-nested parallelized algorithm implementations using 16 processor cores on the test platform and compares these with the scalability issues discussed in Sect. 3.

4.1 Test Setup

In the presented results, the algorithm has been used to calculate all arithmetic feature sets of the input images. The input images are ten images, each containing nine spectral bands in a resolution of 777×776 pixels. The light intensity of each pixel is represented by a double precision floating point number.

The test platform used for producing the results in this paper is a Sun Fire E6900. The machine has 48 UltraSPARC IV CPUs. Each processor has two cores running at 1200 MHz and has 8 MB L2 cache per core. The machine is running Solaris 10. Compilation has been done using the Sun C compiler version 5.9 patch 124867-01 using these options: `-fast -xarch=sparcvis2 -m32 -xopenmp=parallel -lm`.

The image loading time has been excluded from the measurements by loading all ten images, one by one, into main memory before they are processed. Warm up is done by processing all ten images once. The presented results are based on the average execution times of ten or 20 consecutive runs of each feature set, where all ten images are processed. The number of runs is determined by the run time of the particular test case. This is done to increase the accuracy of the measurements, as the processing time is limited by the image size. Using larger input images is not representative for the practical use of the algorithm and will lead to unrealistic results.

The average sequential execution times for feature sets 2 and 3 are 35 s and 127 s respectively, processing all ten multi-spectral image.

4.2 Parallel Efficiency

All tests have been limited to a maximum of 16 processor cores. Several parallelization approaches have been tested to investigate how the two levels of parallelism, l_0 and l_1, influence the parallel efficiency. It should be noted that even though all tests have 16 processors available, they may not all be utilized, depending on the number of threads in the particular test case. The nested version creates more than 16 threads in some tests. In order to prevent the threads to use more than 16 processor cores in these cases, a 16 core processor affinity set was specified using the SUNW_MP_PROCBIND environment variable for all runs with the nested version. This method may potentially lead to uneven load on the cores, but dynamic workload scheduling counters this effect and it is not observable in the results. Even though the main focus of the tests is parallel efficiency, scalability trends can also be extracted from the results of the nested version.

Figures 3 and 4 illustrate the speedup obtained in feature sets 2 and 3 for the nested version by increasing the number of threads at l_0 with different data set partitioning at l_1. As mentioned in Sect. 3.3, one l_1-thread is created for each subset. The measurements of feature set 1, 4 and 5 are not significantly different from what can be observed in feature set 2 and 3, thus they are not shown.

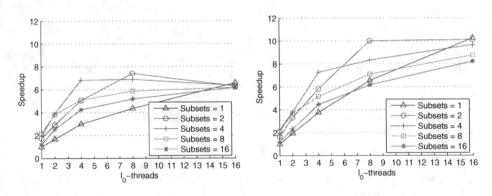

Fig. 3. Speedups for the nested version of feature set 2 with 16 processors

Fig. 4. Speedups for the nested version of feature set 3 with 16 processors

Parallelization at l_0 does not impose any parallelization overhead except for thread creation overhead. However, parallelism is limited to nine l_0 threads. Linear relative speedup should be expected, when more threads can be created to utilize more processors. This can be observed in Fig. 3 for one to eight threads with no data set partitioning for feature set 2, which means $w = 9$. As discussed in Sect. 3.2, going from eight to 16 threads would double the theoretical speedup since load imbalance is improved. However, a speedup of only 1.5 is obtained, because $t_r > t_l$ meaning that data has to be fetched from a remote Uniboard leading to higher memory latency.

This effect has been confirmed by measuring the execution time of each l_0-thread, when running three and nine threads in parallel without any nested l_1-threads. The Sun Fire E6900 UltraSPARC IV Uniboards have four processors each with two cores, which means that if more than eight threads are used, some of them will be running on different processor boards. Figures 5 and 6 show histograms of thread execution time using three and nine threads. It can be seen that using three threads, the histogram is a uniform distribution with a narrow range, while the histogram of nine threads is spread out. The lower part represents threads running on the board that holds the main memory containing the images, while the upper part is slow threads running on a different board. The ratio between a fast and a slow thread match the speedup obtained going from eight to 16 l_0-threads in Fig. 3.

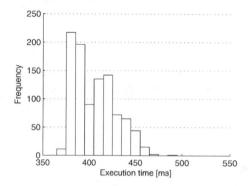

Fig. 5. Thread execution time histogram when running 3 threads

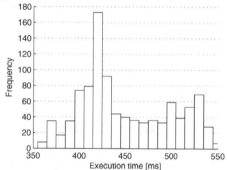

Fig. 6. Thread execution time histogram when running 9 threads

As discussed in Sect. 3.1 parallelization at l_1 has sequential overhead. This can be observed in Figs. 3 and 4 when comparing the speedups of tests with one l_0-thread and increasing the number of l_1 threads. Even though more processors are utilized, the sequential merge eventually outweighs the parallelization speedup. Having more threads than processors also adds thread switching overhead as several threads share a single processor core. It can be observed on both Figs. 3 and 4 that matching $s_0 \times s_1 = |P|$ leads to best results in general.

The effects observed in the results of feature set 2 can also be seen for feature set 3. However, the amount of parallelism available at l_0 is potentially 36 data sets. This leads to better parallel efficiency as less parallelism needs to be extracted at the l_1 level, where the sequential parts are limiting. The efficiency observed in feature set 2 is considered more realistic for real uses of this application, as only a subset of the features is typically needed [7].

The coupling between parallelization and the number of threads is removed in the non-nested version. Splitting the data sets creates more workloads that may lead to better workload balancing among the threads. In Figs. 7 and 8, it can be observed that the non-nested version performs up to 20% better than the nested version. We believe that the performance difference is because the non-nested version will not spawn more threads than processors. This leads to less thread switching overhead and potentially better cache performance.

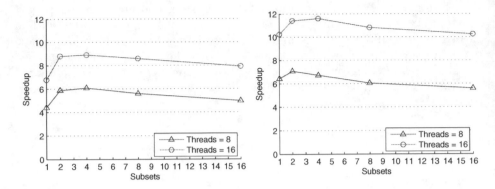

Fig. 7. Speedups for the non-nested version of feature set 2 with 8 and 16 processors

Fig. 8. Speedups for the non-nested version of feature set 3 with 8 and 16 processors

5 Conclusions

This paper has investigated an image processing application that can be targeted for a future multi-processor system-on-chip embedded system. Such a system is inherently parallel, and the major challenges in parallelizing the application have been identified.

These challenges include limited directly exploitable parallelism, a significant parallelization overhead caused by small workloads and difficult load balancing which is aggravated by non-uniform memory latencies.

We have shown that despite these challenges, a relative speedup in excess of 9 on a 16 CPU system can be achieved.

We have experimented with different parallelization approaches. In the results we can observe that a version of the application with nested parallelism is less efficient than a more elaborate version with a single level of parallelism.

Acknowledgements

We gratefully acknowledge the support from the Danish Center for Scientific Computing at the Technical University of Denmark. In particular we acknowledge the technical support from Bernd Dammann. We also would like to thank Jens Sparsø for many thoughtful comments.

References

1. Magarshack, P., Paulin, P.: System-on-chip beyond the nanometer wall. In: Design Automation Conference, Proceedings, pp. 419–424 (2003)
2. Benini, L., De Micheli, G.: Networks on chips: a new SoC paradigm. Computer 35(1), 70–78 (2002)

3. Bertozzi, D., Jalabert, A., Murali, S., Tamhankar, R., Stergiou, S., Benini, L., De Micheli, G.: NoC synthesis flow for customized domain specific multiprocessor systems-on-chip. IEEE Transactions on Parallel and Distributed Systems 16(2), 113–129 (2005)
4. OpenMP Architecture Review Board: OpenMP Application Program Interface 2.5 (2005), http://www.openmp.org
5. Phillips, R., Watson, L., Wynne, R.: Hybrid image classification and parameter selection using a shared memory parallel algorithm. Computers and Geosciences 33(7), 875–897 (2007)
6. Terboven, C., Deselaers, T., Bischof, C., Ney, H.: Shared-memory parallelization for content-based image retrieval. ECCV Workshop on Computation Intensive Methods for Computer Vision, Graz, Austria (May 2006)
7. Clemmensen, L.H., Hansen, M.E., Frisvad, J.C., Ersboll, B.K.: A method for comparison of growth media in objective identification of penicillium based on multispectral imaging. Journal of Microbiological Methods 69(2), 249 (2007)

Scheduling Dynamic OpenMP Applications over Multicore Architectures

François Broquedis, François Diakhaté, Samuel Thibault,
Olivier Aumage, Raymond Namyst, and Pierre-André Wacrenier

INRIA Futurs - LaBRI — Université Bordeaux 1, France

Abstract. Approaching the theoretical performance of hierarchical multicore machines requires a very careful distribution of threads and data among the underlying non-uniform architecture in order to minimize cache misses and NUMA penalties. While it is acknowledged that OpenMP can enhance the quality of thread scheduling on such architectures in a portable way, by transmitting precious information about the affinities between threads and data to the underlying runtime system, most OpenMP runtime systems are actually unable to efficiently support highly irregular, massively parallel applications on NUMA machines.

In this paper, we present a thread scheduling policy suited to the execution of OpenMP programs featuring irregular and massive nested parallelism over hierarchical architectures. Our policy enforces a distribution of threads that maximizes the proximity of threads belonging to the same parallel region, and uses a NUMA-aware work stealing strategy when load balancing is needed. It has been developed as a *plug-in* to the FORESTGOMP OpenMP platform [TBG+07]. We demonstrate the efficiency of our approach with a highly irregular recursive OpenMP program resulting from the generic parallelization of a surface reconstruction application. We achieve a speedup of 14 on a 16-core machine with no application-level optimization.

Keywords: OpenMP, Nested Parallelism, Hierarchical Thread Scheduling, Bubbles, Multi-Core, NUMA, SMP.

1 Introduction

Cache-coherent multiprocessor architectures now commonly introduce multiple levels of locality preference between processor and caches or memory banks. The penalty paid for non-local memory accesses can deeply affect speed-ups when such expensive accesses frequently occur throughout application runs. It is therefore acknowledged that multithreaded programs must carefully distribute threads onto the processors to minimize both cache misses and NUMA penalties. Traditional "opportunistic" scheduling approaches used by most operating systems fail in exploiting hierarchical architectures efficiently however, because they lack information about application behaviour.

Successfully using NUMA architectures requires an in-depth knowledge of the application behaviour in terms of memory access patterns, affinity and

R. Eigenmann and B.R. de Supinski (Eds.): IWOMP 2008, LNCS 5004, pp. 170–180, 2008.

inter-thread collaborations, relationship and synchronization. Gao et al. share this analysis [GSS+06]: They emphasize the importance of exposing domain-specific knowledge semantics to the underlying scheduling layer. Parallel languages such as OpenMP, that rely on the combination of a dedicated compiler and a set of code annotations to extract the parallel structure of applications and to generate scheduling hints for the underlying runtime system, are a great step forward in this respect. However, they currently miss architecture-aware runtime systems that would make an effective and thorough *exploitation* of the gathered knowledge at runtime. As quoted in a proposal for task parallelism in OpenMP [ACD+07]: "The overhead associated with the creation of parallel regions, the varying levels of support in different implementations, the limits to the total number of threads in the application and to the allowed levels of parallelism, and the impossibility of controlling load balancing, make this approach impractical". Moreover, most advanced OpenMP compilers [TTSY00, HD07, THH+05, BS05, DGC05] (featuring super lightweight threads, work stealing techniques, etc.) are not yet NUMA-aware.

In this paper, we present an extension to the GNU OpenMP runtime system that is capable of running dynamic irregular programs over NUMA multicore machines very efficiently. Our runtime generates nested sets of threads called *bubbles*, which encapsulate threads sharing common data, each time an OpenMP parallel region is encountered [TBG+07]. We have designed a NUMA-aware scheduling policy that dynamically maps these *bubbles* onto the various levels of the underlying hierarchical architecture. When load balancing needs to be performed, threads are thus redistributed with respect to their affinity relations. We validate our approach using the OpenMP version of a real-life application (the MPU [OBA+03] parallel surface reconstruction algorithm) that features a highly irregular divide-and-conquer parallel structure based on a recursive refinement process. We show that the OpenMP version of this program clearly draws a substantial benefit from our approach.

2   An OpenMP Platform for Developing and Tuning NUMA-aware Thread Scheduling Policies

To deal with dynamic, irregular OpenMP applications, we claim that the key step is to transmit information extracted by the compiler to the underlying thread scheduler *in a continuous way*. Indeed, only a tight integration of application-provided meta-data and architecture description can let the underlying runtime system take appropriate decisions during the whole application run time.

Thus we have designed "FORESTGOMP ", an extension to the GNU OpenMP runtime system [gom] that relies on the MARCEL/BUBBLESCHED thread scheduling package. BUBBLESCHED provides facilities for attaching various information to groups of threads called *bubbles*, together with a framework that helps to develop schedulers capable of using these metadata.

2.1 Related Work

The numerous studies and papers [MAN⁺99, TTSY00, DSCL04, DGC05, BS05, GSW⁺06, aMST07] that emphasized multilevel parallelism as a promising path toward scalability with OpenMP, gradually brought compiler researchers and vendors to put more of their efforts on OpenMP nested parallelism. Modern OpenMP compilers have some support for nested parallelism and either rely on an efficient user-level thread library (NANOS Nthlib [GOM⁺01], Omni/ST [TTSY00], OMPi [HD07]) or on a pool of threads avoiding useless and costly creation/destruction (Intel Compiler [TGS⁺03, TGBS05], OdinMP [Kar05]).

For instance, Omni/ST is based on a fine-grain thread management system that uses a fixed number of threads to execute an arbitrary number of *filaments*, as with the Cilk multithreaded system [FLR98]. The performance obtained over symmetrical multiprocessors is often very good, mostly because many tasks can be executed sequentially with hardly any overhead when all processors are busy.

To deal with hierarchical architectures, the OMPi C Compiler uses user-level non-preemptive threads that are inserted in the processor runqueues in the following way: threads that are spawned at the first level of parallelism are distributed cyclically and appended at the tail of the ready queues. Inner level threads are inserted at the head of the ready queue of the processor that created them. In order to favor data locality, an idle processor extracts threads from the head of its local queue and steals work from the tail of the remote ones. Moreover the work-stealing scheme follows the computer hierarchy. However, neither Omni/ST nor OMPi provide any support for annotating generated tasks with high level information such as memory affinity. The theft of a thread blindly ignores and breaks the affinity relation between threads that were created together. This may put a strain on the performance on hierarchical, NUMA multiprocessors.

Several OpenMP language extensions have been proposed to control the allocation of work to the participating threads. The mechanism in GOMP [GOM⁺01] to control the binding of threads is useful to tune an application for a given computer. Binding, however, is non-portable from the performance point of view. In order to favor affinities in a more portable manner, the NANOS compiler [DGC05, AGMJ04] allows to associate groups of threads with parallel regions in a static way. The OpenUH Compiler [CHJ⁺06] proposes a mechanism to accurately select the threads of a subteam, although this proposition does not involve nested parallelism.

Finally, the KAI/Intel [STH⁺04] and the NANOS Mercurium compilers [BDG⁺04] support task parallelism and a proposal for parallel tasks in OpenMP 3.0 has been written [ACD⁺07]. This is a major step towards natural support of MIMD applications in OpenMP. Moreover, the OpenMP task paradigm will naturally lead to the generation of structured parallelism, so we claim that the techniques presented in this paper will also be beneficial to programs featuring task parallelism.

3   A Scheduling Policy Guided by Affinity Hints

The challenge of a scheduler for the nested parallelism resides in how to distribute the threads over the machine. This must be done in a way that favors both a good balancing of the computation and, in the case of multi-core and NUMA machines, a good affinity of threads, for better cache effects and avoiding the remote memory access penalty.

3.1   Assumptions

Divide and conquer algorithms generate intensively cooperating groups of threads that run smoother if they are scheduled on the same limited subset of processors. A bad distribution of these collaborating entities results in multiple expensive NUMA accesses over hierarchical architectures, that lowers the general performance of parallel applications. Alternatively, a distribution that considers those affinity relations entails a better use of cache memory, and improves local memory accesses.

The *Affinity* bubble-scheduler is specifically designed to tackle irregular applications based on a divide and conquer scheme. In this aim, we consider that each bubble contains threads and subbubbles that are heavily related, most of the time through data sharing. We assume that the best thread distribution is obtained by scheduling each entity contained in a bubble on the same processor, sometimes breaking the load balancing scheme, even if a local redistribution is needed once in a while. This scheduler provides two main algorithms, to distribute thread and bubble entities over the different processors initially, and to rebalance work if one of them becomes idle.

3.2   Initial Thread Distribution

Entities scheduled in the same bubble should not be torn apart. Nevertheless a bubble can be required to extract its contents to increase the number of executable entities in order to occupy every processor of the architecture. This bubble is then said to be exploded. The runqueue level where a bubble explodes during the distribution is crucial to determine whether affinity relations are preserved or not. For instance, if a bubble is exploded on the top level of the topology, its contents can be scheduled on any processor. Therefore the *Affinity* thread distribution algorithm delays these explosions as much as possible, to maximize locality between the released entities.

More precisely, this first scheduling step is based on a mere recursive algorithm to greedily distribute the hierarchy of bubbles and threads over the hierarchy of runqueues. Upon each call, the algorithm counts the entities available to be distributed from the considered runqueues. If there are enough entities to occupy the complete set of processors covered by the runqueues, entities are greedily distributed over the underlying lists. Otherwise, the algorithm analyzes the contents of each available bubble to determine the ones that hold enough threads or subbubbles to occupy a complete subset of processors on their own. If

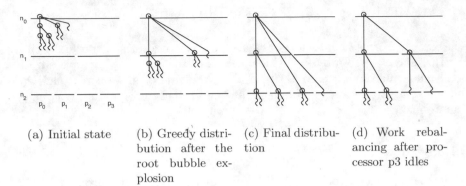

(a) Initial state   (b) Greedy distri-   (c) Final distribu-   (d) Work rebal-
bution after the   tion   ancing after pro-
root bubble ex-   cessor p3 idles
plosion

Fig. 1. Threads and bubbles distribution by the *Affinity* scheduler

so, bubble explosions are delayed to a further step, thus avoiding early separation of collaborating entities.

Figure 1(a) shows the initial state of this recursive algorithm which has been developed to guarantee bubbles and threads distribution from the most general level of the topology, representing the whole computer, to the most specific ones. Figure 1(c) shows the resulting distribution, where only the main top-level bubble has exploded. This approach obviously values affinity relations over load balancing, and could not be efficient without a NUMA-aware work stealing algorithm that rearranges the thread distribution when a processor turns to be inactive.

3.3   NUMA-aware Work Stealing

The irregular behaviour of some applications prevents estimation of the load of each created thread. This lack of load hints forces the *Affinity* scheduler to equally consider every entity. As a result, a continuous thread creation and destruction scheme may unbalance the initial thread repartition, and some of the processors may become idle.

The *Affinity* scheduler implements a dedicated work stealing algorithm to prevent these processors from remaining inactive for too long. This algorithm tracks down lists to steal from, from the most local lists to the most global one if necessary, expanding the search domain as long as no eligible runqueue has been found. Entities are thus stolen as locally as possible. If several entities are usable for work stealing, the *Affinity* scheduler arbitrarily picks the most loaded one, considering the number of recursively contained threads. If only one bubble is found during the stealing process, its contents are browsed to pick a complete subtree of entities, as illustrated by figures 1(d). When a thread, or a bubble, is finally chosen, the algorithm moves its ancestors to the most internal level of the topology common to the source runqueue and the idle processor, to avoid locking convention issues.

3.4 Discussion

A call to the *Affinity*'s thread distribution algorithm generally results in assigning a tree of entities to every processor, similar to the Cilk language or OMPi approaches to deal with divide and conquer applications. Statistically, the working load left to an entity located in the upper part of the tree is bigger than the one executed by the leaves-positioned threads. The *Affinity* scheduler therefore tries to steal from the top of the entity hierarchy, but differs from Cilk implementations by (1) looking for eligible subtrees as close as possible from the idle processor, instead of randomly picking a victim runqueue, and (2) stealing a set of threads that work together rather a lonely thread (like OMPi does). This way of stealing respects the hierarchical nature of both NUMA architectures and the application parallelization scheme.

4 Implicit Surface Reconstruction Application

With *Affinity* and several features of MARCEL, it is now possible to parallelize many recursive divide and conquer algorithms, using a naive approach and simple OpenMP constructs, and yet to obtain good speedups. To back our claim, we show that an extremely irregular divide and conquer algorithm, the Multi-Level Partition of Unity algorithm (MPU) [OBA+03], can be parallelized efficiently only by adding a few lines of code to its implementation.

This surface reconstruction algorithm processes a cloud of points sampling a geometric surface, so as to compute a mathematical representation of this sampled surface. Its main use is related to 3D scanners, that is, devices that are capable of sampling the surface of a physical object by extracting a finite set of 3D points. Reconstructing the whole surface from its samples is required for many applications ranging from rendering to physical simulations.

Thanks to its divide and conquer scheme it is one of the fastest reconstruction algorithms available. Starting from a box containing the whole cloud of points, it tries to fit a simple surface (a quadric) to the points. This surface is implicit, which means that it is defined by a real valued function defined over the entire space and whose value is zero for every point of the implicit surface. If the fitted surface does

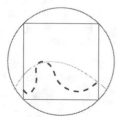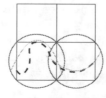

Fig. 2. Adaptive surface fitting using a recursive subdivision of space which forms a tree hierarchy. Each box is subdivided until the fitted surface is close enough to the points. The resulting surface is a weighted average of each local approximation using partition of unity functions.

```
void Node::compute() {
  computeApprox();
  if(_error > _max_error) {
    splitCell();
    for(int i=0; i<8; i++)
      _children[i]->compute();
  }
}
```

```
void Node::compute() {
  computeApprox();
  if(_error > _max_error) {
    splitCell();
#pragma omp parallel for
    for(int i=0; i<8; i++)
      _children[i]->compute();
  }
}
```

Fig. 3. Sequential MPU code to process a node. An approximation is computed and the node is subdvided if it is not precise enough. This process is then repeated recursively.

Fig. 4. Parallel MPU code to process a node. A single OpenMP directive has been added to indicate that every node can be processed concurrently.

not approximate the points closely enough, that is, if there are points too far away from the fitted surface, the box is subdivided into 8 subboxes, thus forming an octree. This process is applied recursively to each child box until the error between each approximation and the points of its box is small enough (see Figure 2).

This divide and conquer approach is made possible by the use of partition of unity functions. Indeed, using these functions makes it possible to define the global reconstructed surface as a weighted average of each function defining the local surface approximations. The weight of each local approximation in the weighted average at a given point in space is at its highest at the center of its box, and decreases as the distance to this center increases.

What makes this technique especially attractive is that there is no "stitching" involved between locally computed surfaces. This makes parallelization easier because such a step would require many synchronisations between threads working on neighbour nodes. Therefore this algorithm is well suited to parallelization since every node of the tree can be processed concurrently. The difficulty resides in balancing the work between the processors as the tree is very irregular and there is no simple way to predict where the tree is going to be refined. Ideally the programmer should be able to simply express that the function calls for processing the nodes can be executed concurrently and the runtime would be responsible for balancing this work on the processors.

OpenMP provides constructs that are very well suited to this task, and parallelizing this algorithm using OpenMP is a matter of inserting a few lines of code to indicate that each time a node is subdivided, its 8 children can be processed concurrently (see Fig. 3 and 4). Running such an application efficiently is challenging for however, because runtime systems need not only to deal with a large number of thread creations/destructions (up to tens of millions for large datasets), but also to schedule them in a way that optimizes memory locality.

5   Evaluation

We validated our approach by experimenting with the MPU application on a cloud of 437644 points, which leads to the creation of 101185 threads.

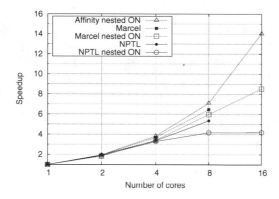

Fig. 5. Speedup of various MPU implementations

The target machine holds 8 dualcore AMD Opteron chips (hence a total of 16 cores) and 64GB of memory. The measured NUMA factor between chips varies from 1.06 (for neighbor chips) to 1.4 (for most distant chips). We tested both the Native POSIX Thread Library of Linux 2.6 (NPTL) and the MARCEL library, partitioning the set of usable cores in order to execute our tests on respectively 2, 4, 8 and 16 cores. The results can be seen on figure 5.

We first tried non-nested approaches to compare the behaviour of these two libraries. Each parallel construct generates a number of threads equal to the number of available cores. The MPU algorithm divides the computed surface in eight different subdomains, every time the refinement primitive is called. Running non-nested tests with a number of threads exceeding 8 is thus not relevant, only the first eight ones will be occupied. The MARCEL thread scheduler operates at user-level, and is less preemptive than the one used by NPTL. MPU thus runs much faster with MARCEL threads.

In the next experiments, we allowed the GOMP compiler to create extra threads when a nested parallel construct is encountered. This approach theoretically suits the MPU application divide and conquer nature. We achieved the best speedups by creating 4 threads at each parallel section. Allowing nested approaches results in creating a great number of threads, and thread creation and management primitives are more expensive in a kernel-level thread library like NPTL. Those used by MARCEL are lighter, which explains why it scales better. On the other hand, neither the runtime system of those libraries has sufficient information about threads' relations to adjust their distribution, so that related threads may be executed by cores located on different NUMA nodes, and the speedup is yet a bit limited. On the contrary, respecting affinity relations by locally scheduling groups of threads results in much better speedups, as can be seen on the *Affinity* curve.

We then evaluated the effectiveness of *Affinity*'s NUMA-aware scheduling algorithm by running two tests. In the first test, the MPU application is *unmodified* but the work stealing algorithm of *Affinity* is replaced by a *random* work stealing algorithm: the victim is elected from a randomly chosen runqueue. The achieved speedup on 16 cores varies between 8.2 and 11.82.

In the second test, the MPU application is *modified*. A single thread is bound per processor, which is used to schedule tasks in the form of lightweight threads. An idle thread tries to steal tasks from the most local queues when necessary. The structure of the application allows this version to use a lock-free stealing strategy most of the time, in a Cilk-like manner. The result is that the Cilk-like version obtains the best speed-up, 15.05 on 16 cores, *at the cost of portability*, since the MPU application was modified to integrate this algorithm. With a speedup of 14.04 out of 16 cores, the FORESTGOMP results almost reach the *cilk-like* results, *without sacrificing* either portability or generality in application-specific optimizations.

6   Conclusion

To exploit nowaday's multiprocessor machines at their full potential, it is crucial to transmit affinity relationships between application threads to the underlying runtime system scheduler. Efficiently scheduling an application on top of a NUMA architecture indeed requires an accurate knowledge of both the machine and the application behaviour in order to make appropriate NUMA-aware scheduling decisions at runtime. Parallel programming languages such as OpenMP are therefore inherently promising since they are particularly fitted for transparent information gathering.

In this paper, we presented a scheduling policy called *Affinity* embedded in our GOMP-based OpenMP scheduling framework and programming environment. *Affinity* is built on the bubble concept and the rich set of manipulation primitives offered by the MARCEL/BUBBLESCHED hierarchical scheduler toolkit to let the application programmer naturally express the thread cooperation affinities and to follow these hints in the actual scheduling process. The experiments we conducted on MPU, a real-life highly irregular surface reconstruction application made a strong case in validating our approach in terms of development *easiness* for the programmer, *portability* and *performance*. Our approach is therefore a way for experts to build complex scheduling strategies that take characteristics of the application into account. Using and mixing such strategies, application programmers get a greater control on scheduling of their OpenMP programs.

In the near future, we intend to investigate two main directions. First, we are currently extending our BUBBLESCHED platform with advanced memory management primitives in order to allocate, register and potentially migrate memory areas used within bubbles on NUMA architectures. This will enable us to take into account memory "attraction" when computing thread redistribution patterns and to operate data movements when significant thread redistributions have to be performed. Second, FORESTGOMP could use static code analysis in determining the groups of threads that are effectively sharing data, and estimating bubble thickness. This information would improve the way the *Affinity* scheduler distributes entities, by naturally preferring the less cooperating threads groups when a bubble must be exploded. This attribute could even be enriched by dynamically refreshed hardware statistics on memory access frequency. Both directions will benefit from our ongoing work towards supporting OpenMP 3.0 tasks in FORESTGOMP.

References

[ACD⁺07]   Ayguade, E., Copty, N., Duranl, A., Hoeflinger, J., Lin, Y., Massaioli, F., Su, E., Unnikrishnan, P., Zhang, G.: A proposal for task parallelism in OpenMP. In: Third International Workshop on OpenMP (IWOMP 2007), Beijing, China (2007)

[AGMJ04]   Ayguade, E., Gonzalez, M., Martorell, X., Jost, G.: Employing Nested OpenMP for the Parallelization of Multi-Zone Computational Fluid Dynamics Applications. In: 18th International Parallel and Distributed Processing Symposium (IPDPS) (2004)

[aMST07]   an Mey, D., Sarholz, S., Terboven, C.: Nested Parallelization with OpenMP. Parallel Computing 35(5), 459–476 (2007)

[BDG⁺04]   Balart, J., Duran, A., Gonzàlez, M., Martorell, X., Ayguadé, E., Labarta, J.: Nanos mercurium: A research compiler for openmp. In: European Workshop on OpenMP (EWOMP) (October 2004)

[BS05]   Blikberg, R., Sørevik, T.: Load balancing and OpenMP implementation of nested parallelism. Parallel Computing, 31(10-12):984–998 (October 2005)

[CHJ⁺06]   Chapman, B.M., Huang, L., Jin, H., Jost, G., de Supinski, B.R.: Extending openmp worksharing directives for multithreading. In: EuroPar 2006 Parallel Processing (2006)

[DGC05]   Duran, A., Gonzàles, M., Corbalán, J.: Automatic Thread Distribution for Nested Parallelism in OpenMP. In: 19th ACM International Conference on Supercomputing, Cambridge, MA, USA, June 2005, pp. 121–130 (2005)

[DSCL04]   Duran, A., Silvera, R., Corbalán, J., Labarta, J.: Runtime adjustment of parallel nested loops. In: Chapman, B.M. (ed.) WOMPAT 2004. LNCS, vol. 3349, Springer, Heidelberg (2005)

[FLR98]   Frigo, M., Leiserson, C.E., Randall, K.H.: The Implementation of the Cilk-5 Multithreaded Language. In: ACM SIGPLAN Conference on Programming Language Design and Implementation (PLDI), Montreal, Canada (June 1998)

[gom]   GOMP – An OpenMP implementation for GCC, http://gcc.gnu.org/projects/gomp/

[GOM⁺01]   Gonzalez, M., Oliver, J., Martorell, X., Ayguade, E., Labarta, J., Navarro, N.: OpenMP Extensions for Thread Groups and Their Run-Time Support. In: Languages and Compilers for Parallel Computing, Springer, Heidelberg (2001)

[GSS⁺06]   Gao, G.R., Sterling, T., Stevens, R., Hereld, M., Zhu, W.: Hierarchical multithreading: programming model and system software. In: 20th International Parallel and Distributed Processing Symposium (IPDPS) (April 2006)

[GSW⁺06]   Gerndt, A., Sarholz, S., Wolter, M., an Mey, D., Bischof, C., Kuhlen, T.: Nested OpenMP for Efficient Computation of 3D Critical Points in Multi-Block CFD Datasets. In: Super Computing (November 2006)

[HD07]   Hadjidoukas, P.E., Dimakopoulos, V.V.: Nested Parallelism in the OMPi OpenMP/C compiler. In: EuroPar, Rennes,France, July 2007, ACM, New York (2007)

[Kar05]   Karlsson, S.: An Introduction to Balder - An OpenMP Run-time Library for Clusters of SMPs. In: International Workshop on OpenMP (IWOMP) (June 2005)

[MAN+99]   Martorell, X., Ayguadé, E., Navarro, N., Corbalán, J., González, M., Labarta, J.: Thread Fork/Join Techniques for Multi-Level Parallelism Exploitation in NUMA Multiprocessors. In: International Conference on SuperComputing, pp. 294–301. ACM Press, New York (1999)

[OBA+03]   Ohtake, Y., Belyaev, A., Alexa, M., Turk, G., Seidel, H.-P.: Multi-level partition of unity implicits. ACM Trans. Graph. 22(3), 463–470 (2003)

[STH+04]   Su, E., Tian, X., Haab, M.G.G., Shah, S., Petersen, P.: Compiler Support of the Workqueuing Execution Model for Intel SMP Architectures. In: European Workshop on OpenMP (EWOMP) (October 2004)

[TBG+07]   Thibault, S., Broquedis, F., Goglin, B., Namyst, R., Wacrenier, P.-A.: An Efficient OpenMP Runtime System for Hierarchical Architectures. In: International Workshop on OpenMP (IWOMP), Beijing,China, June 2007, pp. 148–159 (2007)

[TGBS05]   Tian, X., Girkar, M., Bik, A., Saito, H.: Practical Compiler Techniques on Efficient Multithreaded Code Generation for OpenMP Programs. Comput. J. 48(5), 588–601 (2005)

[TGS+03]   Tian, X., Girkar, M., Shah, S., Armstrong, D., Su, E., Petersen, P.: Compiler and Runtime Support for Running OpenMP Programs on Pentium- and Itanium-Architectures. In: Eighth International Workshop on High-Level Parallel Programming Models and Supportive Environments, April 2003, pp. 47–55 (2003)

[THH+05]   Tian, X., Hoeflinger, J.P., Haab, G., Chen, Y.-K., Girkar, M., Shah, S.: A compiler for exploiting nested parallelism in OpenMP programs. Parallel Comput. 31(10-12), 960–983 (2005)

[TTSY00]   Tanaka, Y., Taura, K., Sato, M., Yonezawa, A.: Performance evaluation of openmp applications with nested parallelism. In: Languages, Compilers, and Run-Time Systems for Scalable Computers, pp. 100–112 (2000)

Visualizing the Program Execution Control Flow of OpenMP Applications*

Karl Fürlinger and Shirley Moore

Innovative Computing Laboratory,
EECS Department,
University of Tennessee, Knoxville
{karl,shirley}@eecs.utk.edu

Abstract. One important aspect of understanding the behavior of an application with respect to its performance, overhead, and scalability characteristics is knowledge of its control flow. In comparison to sequential applications the situation is more complicated in multithreaded parallel programs because each thread defines its own independent control flow. On the other hand, for the most common usage models of OpenMP the threads operate in a largely uniform way, synchronizing frequently at sequence points and diverging only to operate on different data items in worksharing constructs.

This paper presents an approach to capture and visualize the control flow of OpenMP applications in a compact way that does not require a full trace of program execution events but is instead based on a straightforward extension to the data collected by an existing profiling tool.

1 Introduction

An important aspect of understanding the behavior of a parallel application is knowledge about its control flow. In the context of this paper we define the *control flow* as the sequence in which an application executes blocks of code, where a block of code might be as big as a function body or as small as individual statements. Typically, as we will discuss later, in our approach the individual elements of the control flow representations are the source code regions corresponding to whole OpenMP constructs such as parallel regions, critical sections, functions, or user-defined regions. A user can add individual statements to the control flow representation by manually instrumenting them, but typically the user-defined regions would be larger and at least contain a couple of statements.

To motivate the benefit of knowing the control flow of an application, consider the following simple example. Assume our application calls two functions foo() and bar() as show in Fig. 1a. The gprof output corresponding to an execution of this application is shown in Fig. 1c. Now consider the alternative version in Fig. 1b. Analyzing these two applications with gprof gives exactly the same

---

* This work was partially supported by US DOE SCIDAC grant #DE-FC02-06ER25761 (PERI) and NSF grant #07075433 (SDCI).

R. Eigenmann and B.R. de Supinski (Eds.): IWOMP 2008, LNCS 5004, pp. 181–190, 2008.

profile, even though the control flow with respect to the functions `foo()` and `bar()` is different. In the first example `bar()` is always called after `foo()` (20 times) while in the second case `foo()` is the predecessor of `bar()` in the control flow only once (at the beginning of the loop), while it is its own predecessor 19 times. This is visualized in Figs. 1d and 1e, respectively.

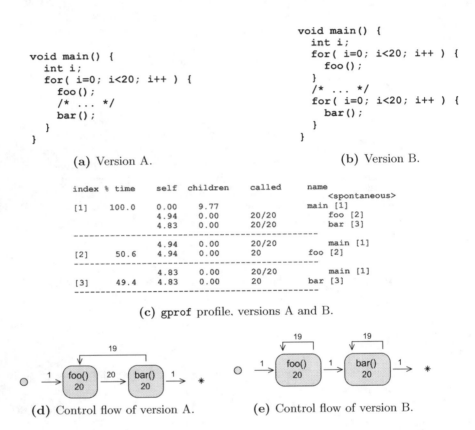

```
       void main() {                   void main() {
         int i;                          int i;
         for( i=0; i<20; i++ ) {         for( i=0; i<20; i++ ) {
           foo();                          foo();
           /* ... */                     }
           bar();                        /* ... */
         }                               for( i=0; i<20; i++ ) {
       }                                   bar();
                                         }
                                       }
```

(a) Version A.   (b) Version B.

```
index % time   self  children  called   name
                                                <spontaneous>
[1]    100.0   0.00    9.77              main [1]
               4.94    0.00     20/20        foo [2]
               4.83    0.00     20/20        bar [3]
----------------------------------------------------------
               4.94    0.00     20/20        main [1]
[2]     50.6   4.94    0.00     20          foo [2]
----------------------------------------------------------
               4.83    0.00     20/20        main [1]
[3]     49.4   4.83    0.00     20          bar [3]
----------------------------------------------------------
```

(c) gprof profile, versions A and B.

(d) Control flow of version A.   (e) Control flow of version B.

Fig. 1. A simple example demonstrating that differences in the control flow are not reflected in runtime profiles

Knowledge about the control flow can be important with respect to performance considerations related to data locality and reuse. If `foo()` and `bar()` work on the same data items, version A keeps data in cache which can be beneficial over version B, which iterates over all data items twice. Evidently the control flow information is not retained in the `gprof` profiles, as in both cases the functions have been called the same number of times and in both cases `bar()` as well as `foo()` have been called from `main()`. Hence, analyzing the callgraph cannot uncover the control flow information.

One approach to recover the control flow is of course to do a full trace of all enter and exit events of all interesting functions, constructs or other source code regions and to visually analyze this trace with tools like Vampir [9], Intel Trace Analyzer [4] or Paraver [10]. However, with raw trace visualization it can be cumbersome to visualize the essential parts of the control flow as the number of events is often overwhelming. In this paper, we discuss an approach that shows that full tracing is not necessary and that the control flow information can be uncovered using a simple extension of a profiling tool.

The rest of this paper is organized as follows: the next section introduces the profiling tool that we extended to extract the control flow information and describes the necessary extensions. Sect. 3 then discusses the visualization of the control flow for OpenMP constructs and presents an example control flow of an application from the NAS parallel benchmark suite. In Sect. 4 we describe related work and in Sect. 5 we conclude and outline directions for future work.

2   The OpenMP Profiler ompP

ompP is a profiling tool for OpenMP applications designed for Unix-like systems. Since it is independent of the OpenMP compiler and runtime system, it works with any OS/compiler combination. ompP differs from other profiling tools like gprof or OProfile [7] in primarily two ways. First, ompP is a measurement based profiler and does not use program counter sampling. The instrumented application invokes ompP monitoring routines that enable a direct observation of program execution events (like entering or exiting a critical section). The direct measurement approach can potentially lead to higher overheads when events are generated very frequently, but this can be avoided by instrumenting such constructs selectively. An advantage of the direct approach is that the results are not subject to sampling inaccuracy and hence they can also be used for correctness testing in certain contexts.

The second difference lies in the way of data collection and representation. While general profilers work on the level of functions, ompP collects and displays performance data in the user model of the execution of OpenMP events [5]. For example, the data reported for critical section contain not only the execution time but also list the time to enter and exit the critical construct (enterT and exitT, respectively) as well as the accumulated time each threads spends inside the critical construct (bodyT) and the number of times each thread enters the construct (execC). An example profile for a critical section is given in Fig. 2

```
R00002 main.c (20-23) (unnamed) CRITICAL
    TID       execT       execC       bodyT       enterT       exitT
      0        1.00           1        1.00         0.00        0.00
      1        3.01           1        1.00         2.00        0.00
      2        2.00           1        1.00         1.00        0.00
      3        4.01           1        1.00         3.01        0.00
    SUM       10.02           4        4.01         6.01        0.00
```

Fig. 2. Profiling data delivered by ompP for a critical section

Profiling data in a similar style is delivered for each OpenMP construct, the columns (execution times and counts) depend on the particular construct. Furthermore, ompP supports the query of hardware performance counters through PAPI [1] and the measured counter values appear as additional columns in the profiles. In addition to OpenMP constructs that are instrumented automatically using Opari [8], a user can mark arbitrary source code regions such as functions or program phases using a manual instrumentation mechanism. Function calls are automatically instrumented on compilers that support this feature (e.g., -finstrument-functions) for the GNU compilers.

Profiling data are displayed by ompP both as flat profiles and as callgraph profiles, giving both inclusive and exclusive times in the latter case. The callgraph profiles are based on the callgraph that is recorded by ompP. An example callgraph is shown in Fig. 3. The callgraph is largely similar to the callgraphs given by other tools, such as callgrind [11], with the exception that the nodes are not only functions but also OpenMP constructs and user-defined regions, and the (runtime) nesting of those constructs is shown in the callgraph view. The callgraph that ompP records is the union of the callgraph of each thread. That is, each node reported has been executed by at least one thread.

```
    ROOT  [critical.i686.ompp: 4 threads]
  REGION  +-R00004 main.c (40-51) ('main')
PARALLEL     +-R00005 main.c (44-48)
  REGION     |-R00001 main.c (20-22) ('foo')
  REGION     |  +-R00002 main.c (27-32) ('bar')
CRITICAL     |     +-R00003 main.c (28-31) (unnamed)
  REGION     +-R00002 main.c (27-32) ('bar')
CRITICAL        +-R00003 main.c (28-31) (unnamed)
```

Fig. 3. Example callgraph view of ompP

2.1   Data Collection to Reconstruct the Control Flow Graph (CFG)

As discussed in the introduction, the callgraph does not contain enough information to reconstruct the CFG. However, a full trace is not necessary either. It is sufficient to keep a record that lists all predecessor nodes and how often the predecessors have been executed for each callgraph node. A predecessor node is either the parent node in the callgraph or a sibling node on the same level. A child node is not considered a predecessor node because the parent–child relationship is already covered by the callgraph representation. An example of this is shown in Fig. 4. The callgraph (lower part of Fig. 4) shows all possible predecessor nodes of node A in the CFG. They are the siblings B and C, and the parent node P. The numbers next to the nodes in Fig. 4 indicate the predecessor nodes and counts after one iteration of the outer loop (left hand side) and at the end of the program execution (right hand side), respectively.

Implementing this scheme in ompP was straightforward. ompP already keeps a pointer to the *current* node of the callgraph (for each thread) and this scheme

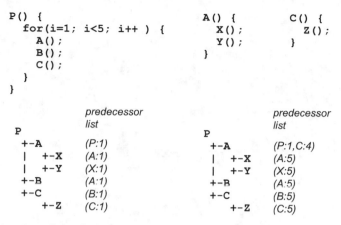

```
P() {                           A() {        C() {
   for(i=1; i<5; i++ ) {          X();         Z();
      A();                        Y();        }
      B();                     }
      C();
   }
}
```

		predecessor list				predecessor list
P			P			
+-A		(P:1)	+-A			(P:1,C:4)
|	+-X	(A:1)	|	+-X		(A:5)
|	+-Y	(X:1)	|	+-Y		(X:5)
+-B		(A:1)	+-B			(A:5)
+-C		(B:1)	+-C			(B:5)
	+-Z	(C:1)		+-Z		(C:5)

Fig. 4. Illustration of the data collection process to reconstruct the control flow graph

is extended by keeping a *previous* node pointer as indicated above. Again this information is kept on a per-thread basis, since each thread can have its own independent callgraph as well as flow of control.

The previous pointer always lags the current pointer one transition. Prior to a parent → child transition, the current pointer points to the parent while the previous pointer either points to the parent's parent or to a child of the parent. The latter case happens when in the previous step a child was entered and exited. In the first case, after the parent → child transition the current pointer points to the child and the previous pointer points to the parent. In the latter case the current pointer is similarly updated, while the prior pointer remains unchanged. This ensures that the previous nodes of siblings are correctly handled.

With current and previous pointers in place, upon entering a node, information about the previous node is added to the list of previous nodes with an execution count of 1, or, if the node is already present in the predecessor list, its count is incremented.

3 Visualizing the CFG of OpenMP Applications

The data generated by ompP's control flow analysis can be displayed in two forms. The first form visualizes the control flow of the whole application, the second is a layer-by-layer approach. The full CFG is useful for smaller applications, but for larger codes it can quickly become too large to comprehend and cause problems for automatic layout mechanisms. An example of an application's full control flow is shown in Fig. 5. The code corresponds to the callgraph of Fig. 3 where the critical section's body contains work for exactly one second.

Rounded boxes represent source code regions. That is, regions corresponding to OpenMP constructs, user-defined regions or automatically instrumented functions. Solid horizontal edges represent the control flow. An edge label like $i|n$ is interpreted as thread i has executed that edge n times. Instead of drawing

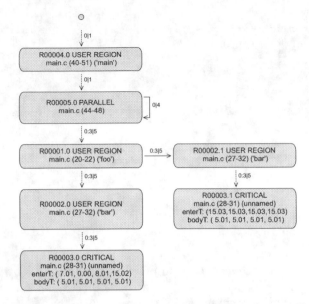

Fig. 5. An example for a full control flow display of an application

each thread's control flow separately, threads with similar behavior are grouped together. For example the edge label 0–3|5 means that threads 0, 1, 2, and 3 combined executed that edge 5 times in total. This greatly reduces the complexity of the control flow graph and makes it easier to understand.

For each node the box contains the most important information. This includes the type of the region (such as CRITICAL), the source code location (file name and line number) and performance data. Due to space limitations the included performance data do not list the full profile but only the most important aspects for the particular construct. This information includes the overall execution time as well as the most likely cause for a potential bottleneck. For critical sections this is the time required to enter the construct (enterT) and for parallel loops it is the waiting time at the implicit barrier, for example.

Dotted vertical lines represent control flow edges from parent to child (with respect to the callgraph). The important difference in interpreting these two types of edges is that a solid edge from A to B means that B was executed after A finished execution while a dotted line from C to D means that D is executed (or called) in the context of C (i.e., C is still "active").

The graphs shown in Figs. 5 and 6 are created with the Graph::Easy tool [2], which takes a textual description of the graph and generates the graph in HTML, SVG, or even ASCII format. For graphs that are not overly complicated the automated layout engine of Graph::Easy does a very good job. However, for bigger graphs a full control flow graph can be unwieldy and it is advisable to do a layer-by-layer visualization in this case.

An example of the layer-by-layer visualization is shown in Fig. 6. Here each graph only shows a single layer of the callgraph, i.e., a parent node and all its child nodes. Since the predecessor nodes of each node are only its siblings or the parent node, this view is sufficient to cover the local view of the control flow graph. The horizontal and vertical edges have the same meaning as in the previous case. To indicate which nodes have child nodes, the text box contains a (+) sign. Clicking on such a node brings up the control flow graph of the child nodes to allow an interactive exploration of the CFG.

The example in Fig. 6 is derived from an execution of the CG benchmark of the OpenMP version of the NAS parallel benchmarks [6] (class C) on a 4-way AMD Opteron processor node (1.8 GHz, 3 GB of main memory). The application is automatically instrumented with Opari and the initialization phase and the iteration loop have been additionally instrumented manually.

As shown in Fig. 6a, the application spends 17.8 seconds in the initialization phase and then executes 75 iterations of the main iteration loop with a total of 702.6 seconds of execution time. Fig. 6b shows the control flow of the initialization phase, while Fig. 6c is the control flow of the main iteration loop. The initialization proceeds in a series of parallel constructs and parallel loops[1]. Significant time is only spent in the regions R00017 and R00027.

Fig. 6c shows the control flow of the iteration loop. We see a nested loop around the R00017 parallel region which is executed 1875 times in total and represents by far the most time consuming region. Region R00017 is called in the initialization as well as in the iteration phase. Drilling down to this parallel region in Fig. 6d, we see that it contains four loops (R00018, R00019, R00020, R00021) of which the first one is the most time consuming. The performance data include the waiting time at the end of worksharing regions (exitBarT). It is an indicator for load imbalance but does show any severe performance problems in this case.

Note that in Figs. 6a, 6b, and 6c the edges are only executed by the master thread (thread 0). Since the application executes sequentially in the phases outside of parallel regions (only the master thread is active). Only after a parallel region is entered, a thread team (with four threads in this case) is created and several threads show up in the control flow graph as in Fig. 6d.

4   Related Work

Control flow graphs are an important topic in the area of code analysis, generation, and optimization. In that context, CFGs are usually constructed based on a compiler's intermediate representation (IR) and are defined as directed multi-graphs with nodes being basic blocks (single entry, single exit) and nodes representing branches that a program execution *may* take (multithreading is hence not directly an issue). The difference to the CFGs in our work is primarily twofold. First, the nodes in our graphs are generally not basic blocks but they are usually larger regions of code containing whole functions. Secondly, the nodes in our graphs record transitions that have actually happened during the execution and also contain a count that shows how often the transition occurred.

---

[1] A parallel loop is one of OpenMP's combined parallel-worksharing constructs.

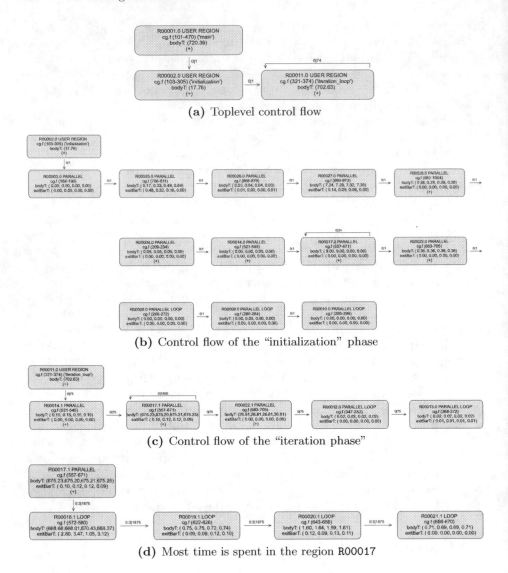

Fig. 6. Four layers of the control flow graph of the CG application of the NAS parallel benchmarks (class C)

Dragon [3] is a performance tool from the OpenUH compiler suite. It can display static as well as dynamic performance data such as the callgraph and control flow graph. The static information is collected from OpenUH's analysis of the source code, while the dynamic information is based on the feedback guided optimization phase of the compiler. In contrast to our approach, the displays are based on the compiler's intermediate representation of source code. The elements of our visualization are the constructs of the user's model of execution to contribute to a high-level understanding of the program execution characteristics.

5   Conclusion

We have presented an approach to visualize the control flow graph of OpenMP applications. We have extended an existing profiling tool to collect the data required for the visualization and used a versatile automated layout tool to generate the graph images.

We believe that the CFG represents valuable information to anyone trying to understand the performance characteristics of an application. Naturally, the author of a code might be very well aware already of their application's control flow and benefit little from the insight ompP's control flow graph can offer. For someone working on a foreign code and especially for big and unfamiliar applications, we believe the CFG view is very helpful to get an understanding of the application's behavior, to understand the observed performance behavior and to identify tuning opportunities.

Future work is planned in several directions. First, ompP cannot currently handle nested parallelism but adding support for this is planned for a future release. Visualizing nested parallelism will pose new challenges when displaying the control flow graph as well. Secondly, we plan to develop an integrated viewer for the profiling data delivered by ompP, eliminating the need for an external graph layout mechanism. Among other graphical displays such as overhead graphs this viewer will also be able to display the control flow graph. We plan to support both the full CFG display as well as the layered approach in an interactive way, i.e., navigating between the nodes of the control flow graph and call graph and linking this information to the detailed profiling data as well as the source code.

References

1. Browne, S., Dongarra, J., Garner, N., Ho, G., Mucci, P.J.: A portable programming interface for performance evaluation on modern processors. Int. J. High Perform. Comput. Appl. 14(3), 189–204 (2000)
2. The graph::easy web page, http://search.cpan.org/~tels/Graph-Easy/
3. Hernandez, O., Liao, C., Chapman, B.: Dragon: A Static and Dynamic Tool for OpenMP. In: Chapman, B.M. (ed.) WOMPAT 2004. LNCS, vol. 3349, pp. 53–66. Springer, Heidelberg (2005)
4. Intel Trace Analyzer, http://www.intel.com/software/products/cluster/tanalyzer/
5. Itzkowitz, M., Mazurov, O., Copty, N., Lin, Y.: An OpenMP runtime API for profiling. In: Accepted by the OpenMP ARB as an official ARB White Paper available online, http://www.compunity.org/futures/omp-api.html
6. Jin, H., Frumkin, M., Yan, J.: The OpenMP implementation of NAS parallel benchmarks and its performance. Technical Report NAS-99-011 (1999)
7. Levon, J.: OProfile, A system-wide profiler for Linux systems. Homepage, http://oprofile.sourceforge.net
8. Mohr, B., Malony, A.D., Shende, S.S., Wolf, F.: Towards a performance tool interface for OpenMP: An approach based on directive rewriting. In: Proceedings of the Third Workshop on OpenMP (EWOMP 2001) (September 2001)

9. Nagel, W.E., Arnold, A., Weber, M., Hoppe, H.-C., Solchenbach, K.: VAMPIR: Visualization and analysis of MPI resources. Supercomputer 12(1), 69–90 (1996)
10. Pillet, V., Labarta, J., Cortes, T., Girona, S.: PARAVER: A tool to visualise and analyze parallel code. In: Proceedings of WoTUG-18: Transputer and Occam Developments, vol. 44, pp. 17–31. IOS Press, Amsterdam (1995)
11. Weidendorfer, J., Kowarschik, M., Trinitis, C.: A Tool Suite for Simulation Based Analysis of Memory Access Behavior. In: Bubak, M., van Albada, G.D., Sloot, P.M.A., Dongarra, J. (eds.) ICCS 2004. LNCS, vol. 3038, pp. 440–447. Springer, Heidelberg (2004)

Author Index

Lecture Notes in Computer Science

Sublibrary 1: Theoretical Computer Science and General Issues

For information about Vols. 1– 4671
please contact your bookseller or Springer

Vol. 4855: V. Arvind, S. Prasad (Eds.), FSTTCS 2007: Foundations of Software Technology and Theoretical Computer Science. XIV, 558 pages. 2007.

Vol. 4854: L. Bougé, M. Forsell, J.L. Träff, A. Streit, W. Ziegler, M. Alexander, S. Childs (Eds.), Euro-Par 2007 Workshops: Parallel Processing. XVII, 236 pages. 2008.

Vol. 4851: S. Boztaş, H.-F.(F.) Lu (Eds.), Applied Algebra, Algebraic Algorithms and Error-Correcting Codes. XII, 368 pages. 2007.

Vol. 4848: M.H. Garzon, H. Yan (Eds.), DNA Computing. XI, 292 pages. 2008.

Vol. 4847: M. Xu, Y. Zhan, J. Cao, Y. Liu (Eds.), Advanced Parallel Processing Technologies. XIX, 767 pages. 2007.

Vol. 4846: I. Cervesato (Ed.), Advances in Computer Science – ASIAN 2007. XI, 313 pages. 2007.

Vol. 4838: T. Masuzawa, S. Tixeuil (Eds.), Stabilization, Safety, and Security of Distributed Systems. XIII, 409 pages. 2007.

Vol. 4835: T. Tokuyama (Ed.), Algorithms and Computation. XVII, 929 pages. 2007.

Vol. 4818: I. Lirkov, S. Margenov, J. Waśniewski (Eds.), Large-Scale Scientific Computing. XIV, 755 pages. 2008.

Vol. 4800: A. Avron, N. Dershowitz, A. Rabinovich (Eds.), Pillars of Computer Science. XXI, 683 pages. 2008.

Vol. 4783: J. Holub, J. Žďárek (Eds.), Implementation and Application of Automata. XIII, 324 pages. 2007.

Vol. 4782: R. Perrott, B.M. Chapman, J. Subhlok, R.F. de Mello, L.T. Yang (Eds.), High Performance Computing and Communications. XIX, 823 pages. 2007.

Vol. 4771: T. Bartz-Beielstein, M.J. Blesa Aguilera, C. Blum, B. Naujoks, A. Roli, G. Rudolph, M. Sampels (Eds.), Hybrid Metaheuristics. X, 202 pages. 2007.

Vol. 4770: V.G. Ganzha, E.W. Mayr, E.V. Vorozhtsov (Eds.), Computer Algebra in Scientific Computing. XIII, 460 pages. 2007.

Vol. 4769: A. Brandstädt, D. Kratsch, H. Müller (Eds.), Graph-Theoretic Concepts in Computer Science. XIII, 341 pages. 2007.

Vol. 4763: J.-F. Raskin, P.S. Thiagarajan (Eds.), Formal Modeling and Analysis of Timed Systems. X, 369 pages. 2007.

Vol. 4759: J. Labarta, K. Joe, T. Sato (Eds.), High-Performance Computing. XV, 524 pages. 2008.

Vol. 4746: A. Bondavalli, F. Brasileiro, S. Rajsbaum (Eds.), Dependable Computing. XV, 239 pages. 2007.

Vol. 4743: P. Thulasiraman, X. He, T.L. Xu, M.K. Denko, R.K. Thulasiram, L.T. Yang (Eds.), Frontiers of High Performance Computing and Networking ISPA 2007 Workshops. XXIX, 536 pages. 2007.

Vol. 4742: I. Stojmenovic, R.K. Thulasiram, L.T. Yang, W. Jia, M. Guo, R.F. de Mello (Eds.), Parallel and Distributed Processing and Applications. XX, 995 pages. 2007.

Vol. 4739: R. Moreno Díaz, F. Pichler, A. Quesada Arencibia (Eds.), Computer Aided Systems Theory – EUROCAST 2007. XIX, 1233 pages. 2007.

Vol. 4736: S. Winter, M. Duckham, L. Kulik, B. Kuipers (Eds.), Spatial Information Theory. XV, 455 pages. 2007.

Vol. 4732: K. Schneider, J. Brandt (Eds.), Theorem Proving in Higher Order Logics. IX, 401 pages. 2007.

Vol. 4731: A. Pelc (Ed.), Distributed Computing. XVI, 510 pages. 2007.

Vol. 4728: S. Bozapalidis, G. Rahonis (Eds.), Algebraic Informatics. VIII, 291 pages. 2007.

Vol. 4726: N. Ziviani, R. Baeza-Yates (Eds.), String Processing and Information Retrieval. XII, 311 pages. 2007.

Vol. 4719: R. Backhouse, J. Gibbons, R. Hinze, J. Jeuring (Eds.), Datatype-Generic Programming. XI, 369 pages. 2007.

Vol. 4711: C.B. Jones, Z. Liu, J. Woodcock (Eds.), Theoretical Aspects of Computing – ICTAC 2007. XI, 483 pages. 2007.

Vol. 4710: C.W. George, Z. Liu, J. Woodcock (Eds.), Domain Modeling and the Duration Calculus. XI, 237 pages. 2007.

Vol. 4708: L. Kučera, A. Kučera (Eds.), Mathematical Foundations of Computer Science 2007. XVIII, 764 pages. 2007.

Vol. 4707: O. Gervasi, M.L. Gavrilova (Eds.), Computational Science and Its Applications – ICCSA 2007, Part III. XXIV, 1205 pages. 2007.

Vol. 4706: O. Gervasi, M.L. Gavrilova (Eds.), Computational Science and Its Applications – ICCSA 2007, Part II. XXIII, 1129 pages. 2007.

Vol. 4705: O. Gervasi, M.L. Gavrilova (Eds.), Computational Science and Its Applications – ICCSA 2007, Part I. XLIV, 1169 pages. 2007.

Vol. 4703: L. Caires, V.T. Vasconcelos (Eds.), CONCUR 2007 – Concurrency Theory. XIII, 507 pages. 2007.

Vol. 4700: C.B. Jones, Z. Liu, J. Woodcock (Eds.), Formal Methods and Hybrid Real-Time Systems. XVI, 539 pages. 2007.

Vol. 4699: B. Kågström, E. Elmroth, J. Dongarra, J. Waśniewski (Eds.), Applied Parallel Computing. XXIX, 1192 pages. 2007.

Vol. 4698: L. Arge, M. Hoffmann, E. Welzl (Eds.), Algorithms – ESA 2007. XV, 769 pages. 2007.

Vol. 4697: L. Choi, Y. Paek, S. Cho (Eds.), Advances in Computer Systems Architecture. XIII, 400 pages. 2007.

Vol. 4688: K. Li, M. Fei, G.W. Irwin, S. Ma (Eds.), Bio-Inspired Computational Intelligence and Applications. XIX, 805 pages. 2007.

Vol. 4684: L. Kang, Y. Liu, S. Zeng (Eds.), Evolvable Systems: From Biology to Hardware. XIV, 446 pages. 2007.

Vol. 4683: L. Kang, Y. Liu, S. Zeng (Eds.), Advances in Computation and Intelligence. XVII, 663 pages. 2007.

Vol. 4681: D.-S. Huang, L. Heutte, M. Loog (Eds.), Advanced Intelligent Computing Theories and Applications. XXVI, 1379 pages. 2007.

Vol. 4672: K. Li, C. Jesshope, H. Jin, J.-L. Gaudiot (Eds.), Network and Parallel Computing. XVIII, 558 pages. 2007.